Computing and Philosophy in Asia

Computing and Philosophy in Asia

Edited by

Soraj Hongladarom

CAMBRIDGE SCHOLARS PUBLISHING

Computing and Philosophy in Asia, edited by Soraj Hongladarom

This book first published 2007 by

Cambridge Scholars Publishing

15 Angerton Gardens, Newcastle, NE5 2JA, UK

British Library Cataloguing in Publication Data
A catalogue record for this book is available from the British Library

ISBN 1-84718-119-8; ISBN 13: 9781847181190

TABLE OF CONTENTS

PREFACE

COMPUTING AND PHILOSOPHY IN ASIA

This volume is a collection of selected papers from the Second Asia-Pacific Computing and Philosophy Conference, which was held at Chulalongkorn University in Bangkok, Thailand from January 7 to 9, 2005. The conference is one of the series of conferences, now being in all major parts of the world, that are organized by the International Association of Computing and Philosophy (www.iacap.org). This conference was the first time that a computer and philosophy conference was held in Asia proper, the first being held in Canberra, Australia in late 2003.

Among the more than 75 papers presented at the conference, 18 are chosen for this volume according to their quality as well as relevance to the theme. One might wonder what is peculiar about Asia that merits a book on computing and philosophy of its own. Does the title "Computing and Philosophy in Asia" mean that this is a book only about computing and philosophy that only happens to be in Asia, or is there anything peculiar to Asia that makes computing and philosophy to be somehow a different kind, perhaps an "Asian" one? This is not an easy question, and in fact both are partly true. On the one hand, the fact that the chapters in this book had their origins as paper presentations in a conference held in Thailand in one sense makes it the case that this is a book on computing and philosophy in Asia. As the world is becoming more integrated, many kinds of activities tend to involve people from different countries; travelling to another part of the world to engage in networking with colleagues from all parts of the globe has become commonplace. In this sense, "computing and philosophy in Asia" presupposes such networking activities. More specifically it presupposes the kind of collaborative action that is the hallmark of academic endeavours since the medieval times (one is reminded of seats of learning such as Paris or Bologna, which were attended by scholars from all over Europe, or the Buddhist Nalanda University in northern India, which was truly cosmopolitan, receiving students from far away places such as China, Korea and Southeast Asia.). According to this way of looking, the content of "computing and philosophy" is truly universal, in much the same way as Christian theology or Buddhist philosophy was regarded as universal in Bologna and Nalanda respectively. That a topic is universal means, of course, that it transcends borders, and it does make no difference whether the persons engaging in its practice hail from France or Italy or Korea. Much of today's scientific enterprise is still true to this ideal.

On the other hand, there is another sense of the title "Computing and Philosophy in Asia" could mean that there is something peculiar about computing and philosophy; here the noun referring to the geographical location becomes an adjective, specifying that there is a difference, not only of the perhaps adventitious geographical location in which an activity takes place, but in its very being. One might talk, of course, of Asian computing and philosophy. The sense being conveyed here is that Asian computing and philosophy is of a different type from, say, the European one, and hence there is nothing universal about the endeavour. An analogy could perhaps be made with food. Among all items that are parts of our daily lives, none is more attached to locality and culture as food. Note that even in this era of globalization in the early twenty-first century, there is still no truly universal or cosmopolitan food, food that could claim allegiance to no particular culture or traditions. The food being sold in fast-food chains all around the world are derived from food in the European tradition. Hamburgers are alien to most Asians, so much so that these fast-food chains have to adapt their menus to suit local tastes in order to compete with the local food stalls (there is an item called "samurai burger" in the MacDonald's chains in Thailand; the burger is made of pork, instead of beef). There is no universal food (thinking about that one seems to have in mind the kind of food that is technologically produced, intended only for nutrition of human beings, and has no taste whatsoever), but Chinese, Japanese, French and Italian food, or food made by one's grandmother.

Interpreted thus, "computing and philosophy in Asia" could mean that there is something unique about it that Europeans might benefit from, in the same way as Europeans have enjoyed Asian food. Perhaps the joining together of "computing" and "philosophy" makes an interesting case here. Computing seems to be a kind of activity that can become truly universal. After all, computing is what is performed on representations of bits and bytes, and as such there seems to be nothing that pertains to a particular cultural tradition. Philosophy, on the other hand, is much involved with culture. Thus one talks of Greek philosophy, Scottish philosophy, Japanese philosophy, and so on. There has been much talk about "universal philosophy," but when analysed this supposedly universal philosophy seems to be derived from one or another existing tradition. The positivist attempt to promulgate a kind of scientific philosophy which does away with metaphysical questions and rigorous verificationist methods appears nowadays to be rather quaint and much tied to its special place and time. Looking back toward history, one finds that the positivists were rather naïve in their belief that philosophical problems could be solved once and for all by eliminating assumptions that were not justifiable through empirical means. Their failure, shown many times over by such philosophers as Sellars, Putnam, Rorty and others, is based, among other things, on the rather simple fact that the key tenet of the positivists itself does not pass their own verificationist justification. This is because the judgment itself is a philosophical one and is not based on any kind of empirical verification or logical analysis. Hence, the idea that philosophy could be based on a foundation that transcends cultures and traditions should, according to its own criterion, be rejected. This shows that the tenet itself is a product of a particular time and place, since it cannot be found on any locality-transcendent foundation at all.

The upshot is that for philosophy itself culture appears to be very much involved. But what about computing? Computer and information technologies are products of a certain kind of society and a certain time and place. They, as are other types of modern technology, evolved as a response to the needs of modern societies, needs which did not happen before in the past. It is also conceivable also that, if the future were very much different from what we are accustomed to right now, the technologies that we are familiar with might not be around then, and the people in the future might look at our contemporary technologies with a curious, distant look. But if this is the case, then computing itself is bound with culture. Nonetheless, talking about "Asian computing" seems rather strange, while "Asian philosophy" is a common term accepted everywhere. Computing as an activity that is much bound up with technology seems to transcend cultures better than philosophy, but since the theoretical foundations of computing are parts of philosophy (as well as much of the work of the computing and philosophy conferences), computing and philosophy are invariably bound up with each other. Hence the fact that philosophy seems to be bound more with cultures than its counterpart is always a problematic one. At least there is a perplexing issue of how these three major concepts—philosophy, computing and culture—are interconnected with one another.

The papers in this volume fully attest to this perplexity. They deal in many different dimensions how culture, computing and philosophy are woven with one another and how many issues and topics in philosophy and related disciplines arise out of these interconnection. Seen in this way, the title, "Computing and Philosophy in Asia" points both to the fact that there could be something substantially different in both computing and in philosophy, when the context in which they are practiced is different from the familiar terrain of the West, as well as that both the computing and the philosophy took place in Asia. We have arranged the chapters in four main groups as follows:

Social and Cultural Concerns

There are four papers dealing with social and ethical issues, as well as with cultures explicitly. The paper by Robert Cavalier deals with e-democracy and the public sphere. Cavalier illustrates how technology, in this case the internet, could be used to enhance the public sphere where citizens in a democracy deliberate on a variety of issues relevant to their communities. In *The Structural Transformation of the Public Sphere*, Jürgen Habermas notes that the role that coffee houses and salons played an important role in the development of a public voice at the intersection of civil society and the state. Such gathering places and the print media that permeated them then became part of the "institutions of the public sphere" that gave rise to an informed "public opinion" in matters of interest to the emerging bourgeois societies of the seventeenth and eighteenth centuries.

It is thus not surprising to see various proponents of "e-democracy" also see connections between internet cafés and a newly emerging internet-based public sphere, and hence there is a profound connection between this public sphere and the very foundations of democracy. Cavalier notes that a critical task for e-democracy is to find ways to model the discursive potential of civil society in such a manner as to provide useful feedback to public policy makers from the local to the national level, and he focuses on an attempt at Carnegie Mellon University in the US which expands both the design and the use of online tools for deliberative polling.

The next paper situates itself in a very different setting. Yeslam Al-Saggaf explores political online forums in his native Saudi Arabia. Faced by strong pressure from the public, the government of Saudi Arabia in 1999 decided to allow its people access to a censored version of the Internet. The reason for the late introduction of the Internet to the country is the Saudi authority's concerns over the arrival of objectionable material, such as pornography and sexually explicit material, in the hands of the public. Most individuals participating in these forums post topics to exchange ideas, discuss their public affairs, and analyse political events. Political online forums have become exceptionally popular in recent years, particularly after the September 11 attacks on the US and the wars against Afghanistan and Iraq. While freedom of expression in Saudi Arabia is limited and content disseminated from traditional media is censored, the decentralised nature of the forums and their many-to-many communication feature make them effective in enabling people to express their views and reach others.

Al-Saggaf finds that television, radio and press are, to some extent, in the hands of the government and content broadcasted from these sources has always been geared towards government's plans. He shows that the forums are the new media that people now have in their hands. Reformists, terrorists, Islamic fundamentalists, moderate liberals and government supporters, are all using these forums to communicate their ideas and plans to others. Their use of political online forums is not only making them authors of media content instead of being a passive audience, but also making these tools a valuable source for information about politics and a perfect medium for people to express themselves and influence the political public sphere in their country.

The next paper takes us across the Indian Ocean from Saudi Arabia to Australia, though its content does not really take us that far away. Aaron Quinn writes in his paper on how technology and journalism play a role in exposing corruption of the press itself. By examining post-September 11, 2001 cases involving Qatar-based satellite news service *Al-Jazirah*, Quinn shows how this news service, once dubbed “the mouthpiece of Osama bin Laden,” may be more balanced than many “free” Western news agencies reporting on the war in Iraq. *Al-Jarirah* has documented numerous violations of human rights, including tenuous issues such as women’s rights, and similar issues often considered taboo in several primarily Islamic nations. Quinn thus illustrates how technology, in this case, helps us identify through comparison who the free press actually is in terms of covering the Iraq war, and in so doing one can then look with a critical eye at the Western press through this juxtaposition. This is a startling conclusion since it seems to be commonly accepted that the Western press is the epitome of professional journalistic standard. Quinn thoroughly examines and criticizes the Western press that, according to him, not only commissions obvious unethical acts, but also, either directly or not, omits duties central to their profession. Here one finds that “computing and philosophy” takes a very critical turn, and looking what the people at one part of the world are doing might well enable those in other parts realize their mistakes which are perhaps hidden by their self serving attitude.

In her paper on the “computerized generation” in Thailand, Chulanee Thianthai explores the lives of Thai teenagers who are wired to modern gadgets such as mobile phones and the Internet. Thianthai presents a clear glimpse into the lives of Thai teenagers whose lives are so permeated with technology that it is hard to distinguish them from their counterparts, say, in the US or Europe. Thianthai catalogues the daily lives of Thai teenagers from waking up until going to bed, and she finds that, although on the surface the lives of these teenagers do not look different from their counterparts in the West, there are indeed elements of “hybridization” in their lives, and here her paper resonates with Ess’ in this volume. Their cultural influences are still there, only that they are modified by the technology. Thianthai’s work is thus an interesting case study for those who would like to know more about how technology and culture interact with each other.

Ethical and Religious Perspectives

The next group of papers concern ethical and religious perspectives as they pertain to computing. Charles Ess' paper is a philosophical attempt to found a kind of ethical pluralism that does service both to the diverse cultures as well as the need for shared judgments. According to Ess, "while these pluralisms can be seen in emerging ethical guidelines for research that straddle considerable cultural differences … , we find that the even greater cultural and ethical differences marking Western and Eastern approaches—most centrally, to issues of privacy and data privacy protection—nonetheless likewise allow for an ethical *pluralism* that holds together important ethical norms alongside the irreducible ethical traditions and approaches that define diverse cultural identities around the globe."

As in the case of the empirical study by Thianthai, Ess argues for a way for a reconciliation of cultural differences and shared norms and judgments, which is a kind of hybridization. While there are sharp differences along cultural lines, for example in terms of how data protection should be understood and enacted, nonetheless there are agreements on such broad terms as the need for protecting the individuals in society. What we are seeing here is then a critique of dichotomous thinking that seems to bifurcate between ethical particularism and universalism. Such a bifurcation, according to Ess, rests on a mistaken assumption that ethical judgments are founded upon a system of justification that belongs to one particular cultural tradition only. With such a narrow viewpoint, it becomes conceivable how one might judge the traditions of others to be alien and perhaps to be inferior to one's own. Ess spends considerable time in the paper detailing the Confucian ethical system of judgment in order to show how a system that looks foreign to most Westerners could yield insights which, while clearly different from that of the West, nonetheless does function just as effectively as a system of computer ethics in today's world. Hence Ess has made an important contribution in the understanding of how the diverse cultural traditions could make a contribution to the global attempt at providing a normative foundation for our computer age.

Remaining with the spiritual traditions of the East, Ken Herold goes to the teaching of Buddhism to find out what is "an informational person." Herold is offering a Buddhist theory of information where information is a construct and is not there objectively in reality. According to Herold, what he calls "informational entity" includes both persons and non-persons. What this means is that both subjects and objects (both are only results of construction and the subject-object distinction does not exist "out there" according to Herold's interpretation of Buddhism) are constituted by information. An upshot of this is, then, that the person herself is an informational entity, which when realized by the person herself leads toward the release from suffering, which is the supreme goal of Buddhism. From his Herold argues that the informational person is the ethical person.

Perhaps we could understand Herold's position better by starting from the basic Buddhist teaching on the nature of reality. According to Buddhism, things do not exist inherently; instead their existence does depend on cognition in an important way. Since existence depends on cognition, things that are cognized as existing are thus informational. This includes persons and selves too. Realizing this, one experiences a kind of release from the hold one has on things and persons that, according to the Buddhist teaching, are the source of suffering. This, for Herold, is a way toward an "onto-ethics" where ontology and ethics are fused together.

The next paper takes us back to Europe and the familiar tradition of the West. In one of the two papers in the volume, Lorenzo Magnani argues for an ethics which is contrary to the Kantian spirit in that it takes into consideration decision making behaviours of human beings. The Kantian spirit, one may recall, is such that there is no need for either science and philosophy in order to become morally good. Knowledge, whether practical or theoretical, has to relevance to ethics according to Kant. On the contrary, Magnani is proposing that ethics and decision-making behaviour should continually and carefully be accompanied by knowledge that is related to an understanding of various problems and situations. In this sense knowledge is considered to be a duty, and hence the problem of how the knowledge is disseminated and distributed becomes morally significant. The knowledge Magnani is proposing is that of model-based rational structure that represents how human beings think ethically and he argues how this type of knowledge contributes to the problem of ethical evaluation. Moreover, he also touches upon the practical problem of privacy and identity in today's cyberworld.

In a reflective paper, John Weckert asks what a good life is for the technological society. The key question in his paper is whether there is such a thing as good life as enhanced by computers. Underlying this question is the question whether technology has any role to play in good life. Starting from Aristotle's view on *eudaimonia,* good life as defined through a metaphysical assumption of the human *telos*, or the end that defines what a human being actually is, Weckert goes on to reflect upon the effects brought about by computer technology on the environment of human beings and our very beings. Computer technology can enhance the lives of us human beings in various ways; for example Weckert asks what we would do if we had the vision of an eagle, and more poignantly, what would happen if only some human beings were so enhanced while others were not. This points to a special situation in which enhanced and unenhanced human beings would become so different from each other that conceptions as to what constitutes good life and happiness for one group may not be fully applicable to the other. Weckert leaves us with the question of what is the point of technological development if enhancing the very being of human beings lead to a radically new way of figuring out what is happiness and good life. It would seem to be a different thing all together.

Metaphysical and Epistemological Perspectives

The next group of selected papers deal with the core issues in philosophy, namely reality and knowledge. Larry Hinman explores how to build a robot that understand Kant's Transcendental Turn. Never mind the fact that there are very few human beings who actually understand Kant's writings, let alone robots, Hinman's point is that building such a robot is an effective thought experiment in getting students to understand what Kant is really up to in his notoriously difficult epistemology. Hinman shows that the problems faced by Kant in the first half of the first *Critique* is not unlike that of a computer scientist in designing a robot. Understanding the basic steps in designing a robot can help students in understanding the Transcendental Analytic and the Transcendental Deduction of the Categories in the first *Critique*. All incoming visual data needs to be stamped with a time and place stamp if they are to be meaningful, or, to put it in Kant's language, space and time are a priori forms of intuition. Furthermore, individual pieces of data then must be related to one another according to certain basic rules. For example, there must be some basic rule that various pictures of an object are pictures of the same object; in Kant's words, the concept of a physical object is an a priori category of the understanding and a necessary condition of the possibility of any meaningful experience at all.

Hinman's point is more than pedagogical, however. Thinking about designing an elementary robot can help us understand Kant's transcendental turn. In so doing, we can begin to understand Kant as a precursor of the computational turn that has characterized recent work in information philosophy. Moreover, thinking about Kant's transcendental turn in terms of elementary robotics allows us to understand the magnitude of the challenge that faced Kant: Imagine if we were robots, how could we discover and prove the validity of our own a priori structures of sensation and understanding?

In his other paper in the volume, Lorenzo Magnani discusses computation abduction as an epistemic mediator and the extra-theoretical dimension of scientific creativity. The common views associate to creativity unusual and mysterious qualities that drive the concept of creativity to a confused verbosity. This conclusion has also been supported by many philosophers of science who studied conceptual change in science during the second half of the last century. Some of them argued that a logic of discovery (and a rational model of discovery) could not exist; scientific conceptual change is cataclysmic and sometimes irrational. Magnani is referring, of course, to Thomas Kuhn's tremendously influential view of scientific change. Against this, Magnani argues that we can overcome many of the difficulties of creativity studies and develop a theory of abduction, in the light of the works of Charles Sanders Peirce.

Magnani maintains that the "computational turn" gives us a new way to understand creative processes in a strictly pragmatic sense. Conceptual tools from Artificial Intelligence and Cognitive Science do allow us to test concepts and ideas previously conceived in abstract terms. It is through the perspective of these actual models that we find the central role of abduction in the explanation of creative reasoning in science. What is called "theoretical abduction" illustrates much of what is important in abductive reasoning, especially the objective of selecting and creating a set of hypotheses that are able to dispense good (preferred) explanations of data. However, it fails to account for many cases of explanation occurring in science or in everyday reasoning when the exploitation of the environment is crucial. Magnani then devises the concept of "manipulative abduction" which captures the role of action in many interesting situations, action that provides otherwise unavailable information that enables the agent to solve problems by starting and performing a suitable abductive process of generation or selection of hypotheses. Many external things, usually inert from the epistemological point of view, can be transformed into "epistemic mediators," namely objects in the external world that serve to facilitate cognitive tasks.

Returning to the theme of human enhancement, Albert Gatti talks about sensing and thinking through technological tools. Having working in an Italian hospital, Gatti shows how cognitive processes are distributed between the individual agent and her external environment, as well as that the physical interaction between the body and the environment plays a fundamental cognitive role. Hence, Gatti agrees with Magnani in that the external environment does play a constitutive role in epistemological and cognitive tasks. The knower is no longer held to be a solitary mind reaching out toward the object, but instead performs many interactive activities in order to gain knowledge.

Taking a case study of laparoscopic video-surgery in the hospital, Gatti explores how this theoretical account plays out in reality. In such a surgery, the surgeon's bodily relation with her patient complete changes, and Gatti shows how the change effects the surgeon's cognitive processes.

In "Growing Minds, Computability, and the Potentially Infinite," Darren Abramson argues for the existence of a set that is neither finite nor infinite. Such a set, according to Abramson, is "potentially infinite." At no point can one traverse all members of such a set, yet it is always possible to traverse more members at any given moment. The idea is that the mind has the capability of contemplating the infinite. Not the actual infinity, which is patently impossible, but the potentially infinite in the sense that the mind always has the capability of extending its store of memory or the time it needs to compute a function, even though at any given time it is limited to finite amount of time or numbers. The upshot of Abramson's idea is that, about the only way to avoid Kripkean finitism, a rather well known skeptical idea derived from his interpretation of Wittgenstein that computing any finite amount of natural numbers is always open to wild interpretations, is to accept that the mind can comprehend the potentially infinite in this way. Here "growing minds" refer to this capability of the mind to grow and to comprehend infinity.

Philosophical Issues in Programming

The next group of papers discusses philosophical and foundational issues in computer programming. Gerald Khoury and Simeon Simoff discusses some philosophical foundations for a unified enterprise modelling language. Khoury and Simoff contend that the ability to create effective Enterprise Architectures (EA's) rests on the ability to create "simple models that are easily communicated … ." However, current EA's are complex, difficult to produce and hard to understand, because EA's are produced using a variety of heterogeneous modelling languages and because there is no widely accepted language that can be used to model all aspects of an enterprise. The development of a unified EA modelling language, hence, is a highly coveted goal. What Khoury and Simoff attempt in the paper is to lay the groundwork for an understanding of organization and system structures that is couched within the framework of philosophic, linguistic and cognitive theories. This understanding is used as the basis for a novel approach to developing a unified EA modelling language. Key to their idea is that there is no difference between a model and a metaphor. As comparison of properties makes sense only in reference to a more abstract, higher level type of property, this higher level is then termed "supertypes." Enterprise models then include supertypes, as they themselves are viewed as metaphors. Thus a way is proposed in which a philosophical unified modelling language.

The next paper takes us back to the enhancement of human beings. Ludmil Duridanov and Simeon Simoff explore the hearing dimension of three-dimensional world. Sometimes old ideas in the history of civilization could be applied in computing technology in the appropriate moment with the aim to have a new impact. Virtual worlds are not a new concept. Since the early days of its known history, humankind has sought to shape its environment. In the process of modelling their physical world, humans tend to create alternate worlds, which are subject to the rules of the human creators—even pretending to be a sacred space "created by God." Those "philosophical" virtual worlds exist only in the minds of their originators (hence the name "virtual"). Technology-mediated virtual worlds are manifested and shared with others through a particular medium "in mind." These worlds can be designed to be experienced through more than one medium.

Duridanov and Simoff then consider computer-mediated virtual worlds. Current emphasis in the representation construction in these virtual worlds is mostly on the richness and efficiency of information visualisation. World developers and designers do not pay enough attention on their audio counterpart. A holistic approach is discussed that moves toward the design of virtual worlds integrating the sound dimension in the virtual architecture of cyberspace. In summary, the work extends the dominating principle "What you see is what you get" to "What you hear and see is where you really live."

The last paper in the group is an example of a fine mixture of conceptual tools from the humanities, that of deconstruction derived from Jacques Derrida, to help with solving problems in computer science. Derrida discusses the notion of program as a form of writing. Writing is essentially a process of "scratching marks" on a physical space and medium. In this regard, the computer is no longer simply a calculating machine; rather, it has become a medium for human communication, a writing space.

One of the most distinctive functions of writing is separation process. The foremost factor of writing in terms of separation is deemed to be the division between the known and the knower. But separation can also be seen not just as simple and static division, but as a process that also joins. Writing is what Derrida's concept of *"différance"* deals with. The motion and fluidity of writing as *différance* is equivalent to the denial of static separation and represents the affirmation of separation as temporary, active difference which seeks to join up with what it has been separated from. Following Derridean logic, Jang applies several deconstruction analyses to computer programming. Computer programming as "electronic writing" institutes its own kind of *différance* in the way in which the computer is able to make its own differences. Taking computer programming as a genre of writing promotes the possibility of deconstructive practices more readily than does the traditional writing of graphic marks. In conclusion, computer programming can be said to be the same process of deconstruction.

Issues in Teaching

We end the volume by offering three papers related to computers and teaching, beginning with a paper on Tiki by Michael Byron. Byron worked on implementing technological enhancements of the traditional pedagogical model, called TikiWiki, or Tiki for short. Byron used this software to create an entire website environment for his philosophy classes, and this environment has allowed his students to learn the course content in ways quite different from the traditional model. The learning is different because (as Dewey emphasized) we learn what we do. Byron describes the Tiki, presents some considerations in favour of deploying it in philosophy courses of very different kinds, and explains how the software is used in classroom settings.

Pannee Suanpang discusses her students' online learning experience in a Thai university. She explores students' experiences in learning in an online course on business statistics course and examines the qualitative reflections that they provided throughout the course and as part of the course evaluation. The research project compared online learning with traditional teaching conducted over sixteen weeks. Qualitative methods used in this study included in-depth interviews, student diaries, messages from the discussion boards and comments on course appraisals. Her findings indicated that the majority of the students used the online mode for reading course materials, doing exercises, searching for information, taking quizzes, and communicating with other students and online instructors. Such an online course can support and assist them to learn more effectively, and enhance their attitude to business statistics. Online students would benefit from increased access to technology in order to assist their learning. The results of this study will be used to develop more effective online learning systems.

Last, but not least, Hulpke and To begin their paper by asking the age-old philosophical question: Can ethics be taught? Put another way, how can students be led to think about ethics, especially in Asia where memorisation is favored over thought? Next question they consider is: How can sense of community, team learning, be fostered in big classes? Finally, how can ideas of individual participation be encouraged in Asia, especially in big classes? As in many places around Asia our university faced exactly these challenges and turned to computer technology for answers. To assess results it may be helpful to first review some frameworks to assess teaching ethics. Then a look at our classroom experiences may give insights into the three questions above.

Conclusion

How does one conclude such a diverse group of papers? Perhaps first by recognizing its diversity and admitting that any kind of overall summary of the whole is closely impossible any way. Nonetheless, one can get a glimpse of what is being offered here. On the one hand, one finds papers that deal specifically with cultural issues in computing as well as with reflections on the underlying foundational and philosophical issues involved. On the other hand, we also have more technical papers that deal with matters in logic and set theory, as well as practical ones that focus on, say, teaching university students and on how the media should reform themselves. Be that as it may, I believe one thread that ties these papers together is that they, collectively, show that cultural dimensions in computing and philosophy are an important factor, one that cannot be ignored in scholarly pursuit in the field. Of course the term "culture" itself is notoriously elusive to pin down, but that only adds to its attractiveness and its fertility in fostering novel interpretations and perspectives that have sustained philosophical investigation and made it flourish.

Soraj Hongladarom
Center for Ethics of Science and Technology
Faculty of Arts
Chulalongkorn University
November 2006

ACKNOWLEDGMENTS

This book would not have been possible if not for the generous help and support that was given me throughout the long process of producing it. First of all I would like to thank all the participants of the Second Asia-Pacific Computing and Philosophy Conference. Their enthusiasm and commitment to excellence in the field of computing and philosophy, not to mention their dedication in traveling to Thailand, which for many lies half way around the globe, deserves all the credit and gratitude. They were instrumental in making the three-day meeting at the Novotel Hotel in the middle of Bangkok a memorable one. Many of the participants have become authors of the chapters in this volume, and my thanks to them for their belief and their patience in waiting a long time to see this volume finally out in print.

Chulalongkorn University has also been giving tremendous support in setting the Center for Ethics of Science and Technology, which is a virtual center for all areas related to science and society and ethical considerations of science and technology. As an interdisciplinary center, coordinating all of its activities is always a challenge, and the conference and this book did also reflect that. I also owe a great deal to the International Association of Computing and Philosophy, especially to Robert Cavalier, who first cajoled me into holding the second AP-CAP conference when we met in Canberra. I learned a lot from him, especially his dedication to computing and philosophy and his zeal in seeing that it is spread to all corners of the world. This is by no means an easy task. The conference was also generously supported by the Australian Center for Applied Philosophy and Public Ethics (CAPPE). John Weckert was the key person for the securing a grant from the CAPPE that enabled two of the keynote speakers at the conference, John Weckert himself and Jeroen van den Hoven, to come to Bangkok. The Regional Unit for Social and Human Sciences, UNESCO also generously provided more grant that enabled many participants in the region to attend the conference. To them I owe a lot of gratitude, especially to Darryl Macer, its Director. Charles Ess has always been a tremendous help and support to me during these past years, and I owe him a great deal. His coming to Thailand was enabled by a grant from the Royal Golden Jubilee Fund that he collaborated with me. The conference was actually part of the research project on "Ethics of the Global Digital Divide," which was supported by the Thailand Research Fund. I would like to thank the Fund so much for this and for continually supporting me and my research group throughout these years.

Finally many of my friends and students at Chula helped a lot with the conference. I would like to thank Krisana Kitiyadisai, Chayunt Mathon and many others who helped with many aspects of the conference.

CHAPTER ONE

E-DEMOCRACY AND THE STRUCTURAL TRANSFORMATION OF THE PUBLIC SPHERE: CARNEGIE MELLON'S PROJECT PICOLA (PUBLIC INFORMED CITIZEN ONLINE ASSEMBLY)

ROBERT CAVALIER
DEPARTMENT OF PHILOSOPHY, CARNEGIE MELLON UNIVERSITY

The Conversational Turn in Political Philosophy

Liberal democracies of the kind we see forming around the world are only the beginning of what Benjamin Barber has called strong democracies (Barber 1984). Thin, liberal democracies provide the constitutional essentials of universal suffrage, freedom of press and assembly, etc. but this in no way guarantees that the citizens of these societies will see themselves as any more than isolated individuals who periodically vote (if they choose to do so). Recent work in "Citizenship Theory" has made clear that "the health and stability of a modern democracy depends, not only on the justice of its basic institutions, but also on the qualities and attitudes of its citizens" (Kymlicka 2002, 285). These qualities and attitudes are often highlighted by proponents of Deliberative Democracy and emphasize the role of the citizen in becoming a truly informed and engaged individual, a person willing to listen to all sides and willing to let the force of the better argument (in all its richness) become a guide to opinion formation.

Over the past 20 years, this concept of Deliberative Democracy has emerged as a major contribution to democratic theory. With philosophical foundations in the later writings of Jürgen Habermas and John Rawls, the concept has also received considerable attention within the social sciences.[1]

---

[1] See Rawls "Justice as Fairness: Political not Metaphysical" and "Constitutional Liberty and the Concept of Justice" in *John Rawls: Collected Papers,* ed. Samuel Freeman, *Political Liberalism,* and *Public Reason Reconsidered.* See Habermas *Between Facts and*

Put briefly, the citizens of the modern liberal state manifest an irreducible plurality of visions of the good that encompass all but the most basic principles of justice. In turn, the principles of justice that make up the "constitutional essentials" ensuring "life, liberty and the pursuit of happiness" provide the political horizon wherein the diverse visions of the good can be pursued. Much of the work of Habermas and Rawls is devoted to defending the rational foundation of liberal constitutional democracies through arguments revolving around notions of "normative validity" and "overlapping consensus." Fundamental to both of their arguments is what Habermas calls "Dialogical Reason." In this view, Practical Reason moves through conversations characterized by the fact of plurality and the richness of perspectives. Ideally, it reaches consensus on "constitutional essentials" through the uncoerced conversation of rational beings pursuing the stronger argument or the overlapping interests of diverse populations in their desire to live their lives in the light of their own comprehensive worldviews.[2] We see these elements of deliberative democracy in the recent creation of constitutions for the Republics of Afghanistan and Iraq.

"Open and informed" conversations on the part of the citizenry are essential to the concept of deliberative democracy. "Openness" refers to the ability of all perspectives to be allowed a voice in the discussion and "informed" refers to the need for the discussion to be based upon the best information and arguments

Norms: Contributions to a Discourse of Law and Democracy; and "Citizenship and National Identity: Some Reflections on the Future of Europe" (Praxis International) (cf. quote from Kymlicke, p. 285). The relation of deliberative democratic theory to the social sciences has often taken the form of a debate with standard rational choice and social choice approaches to democratic theory: See *The State of Democratic Theory* (Sharpiro). The classic works in social choice theory are Anthony Downs's *An Economic Theory of Democracy* and William Riker's *Liberalism Against Popularlism.* A social history of the evolution of social choice theory and its connection to democratic theory can be found in S. M. Amadae's *Rationalizing Capitalist Democracy: The Cold War Origins of Rational Choice Liberalism.*

[2] Borrowing heavily from the later Rawls and Habermas, I view the citizen as a complex person capable of being (1) "rational" in the sense of being able to use means/ends reasoning in a self-interested way and (2) "reasonable" in the sense of being able to compromise on such self-interested calculations if it seems that others may have alternate views that seem worth supporting or at least acknowledging. For example, I may take a higher rate of property tax if you convince me that an extended bike trail may bring a good quality of life to others even if I don't benefit from these trails or I may assent to the policies that others adopt even though I opposed them with my vote. Furthermore, I follow Rawls in ascribing to this person two "moral powers:" (1) a sense of reciprocity/fairness (empathy) and (2) the capability of generating a vision of the good. The latter accounts for a significant part of "pluralism" and can be seen, for example, in the different political platforms of the American Democratic and Republican parties.

available. These broad requirements of practical reason have been adopted by advocates of deliberative democracy with the goal of achieving a strategy for democratic decision-making on a range of practical political problems.[3]

While there is considerable debate within the literature over the appropriate scope and epistemological nature of these conversations, there is general agreement about the Deweyan belief in social intelligence and the hope that we "think best when we think together."[4] Indeed, unless you are an anti-democrat like Karl Popper's Plato, you would hope that the "voice of the people" (*vox populi*) could be brought to bear on policy decisions that affect the people. And it is here that the role of "public opinion" enters the discussion.

The Rise of Modern Public Opinion

The age-old argument against public opinion (*doxa*) goes back to Plato's critique of the Sophists and his distinction in the *Crito* and elsewhere between "true" opinion and "false" opinion. The distinction itself and the path toward true belief are characterized by philosophical dialectic – a practice developed and refined in the education of the philosopher. For surely it does not follow from the fact that "everyone says the earth is flat," that the earth indeed is flat. Anyone arguing in this manner—in court or in private—is simply committing the Ad Populum Fallacy.

But these ancient distinctions conceal as much as they reveal. The history of "public opinion" is far richer and more complex than the Ad Populum fallacy might imply. In his work on *The Structural Transformation of the Public Sphere*, Habermas describes the rise of a more forceful meaning of public opinion in the bourgeois societies of the 17th and 18th centuries. Here the appearance of the press and its demands for more freedom along with the growth of coffee houses and salons gave rise to a new "public sphere" capable of questioning and arguing with the established State and its aristocratic trappings. (Habermas 1989, 14-34) Throughout the 18th Century this new Civil Society developed its own voice, which first found expression in newly coined terms such as "*opinion publique*" (Habermas 1989, 95).

---

[3] See Samuel Freeman, "Deliberative Democracy: A Sympathetic Comment" (*Philosophy and Public Affairs* (29, No. 4) for an overview of perspectives on deliberative democracy and Bohman and Rehig *Deliberative Democracy* along with *Debating Deliberative Democracy* ed Fishkin for more individual articles and essays.

[4] See James Campbell's *Understanding John Dewey* "The process of living in democratic community requires a recognition that our political life "is essentially a cooperative undertaking, one which rests upon persuasion, upon ability to convince and be convinced by reason; or, in ordinary language, upon public opinion' (p. 200; the quote from Dewey is from his *Democracy and Education* Middle Works 10, p. 404).

As the sphere of the political expanded to include more and more of "the people," so too did the need to listen to that voice, at least the voice cultivated by the information and discussions of this new, bourgeois civil society. In analyzing a passage from Edmund Burke, Habermas writes: "The opinion of the public that *put its reason to use* was no longer just opinion: it did not arise from mere inclination but from private reflection upon public affairs and from their public discussion" (Habermas 94, my emphasis).[5] This opinion, Burke and many others during this period noted, should and could not be ignored.

On the one hand, public opinion could be the source of "wisdom and sagacity" and, on the other, a source of check and balance.[6] Please provide complete reference for Bentham. On both counts, it is now perceived to be essential to the proper running of government. A government, moreover, that now had to make its own deliberations public. An order of the Prussian King in 1804 proclaimed "that a decent publicity is for both government and subjects the surest guarantee against the negligence and spite of subaltern officials and deserves to be promoted and protected by all means" (Habermas 1989, 84). Related to these proto-"sunshine Laws" were constitutional movements pointing the way toward greater suffrage and freedom of expression. Regarding the former, Madison wrote "who are the electors of the federal representatives? Not the rich, more than the poor; not the learned, more than the ignorant …The electors are the great body of the people of the United States" (Hamilton 2004, 57); for the latter, see the *Declaration des Droits de l'Homme et du Citoyen* "The right to communicate one's ideas and opinions, whether through the press or in any other manner, the right to assemble peaceably… cannot be refused" (quoted in Habermas 1989, 70).

Correlating these constitutional guarantees with social decision making, Fox, in a 1792 speech before the House of Commons, said: "It is certainly right and prudent to consult the public opinion…if the public opinion did not happen to square with mine; if, after pointing out to them the danger, they did not see it in the same light with me … I should consider it my due to the king … to retire, that they might pursue the plan that they thought better, by a fit instrument, that is by a man who thought with them…" (Habermas 1989, 66). Today, we call that "fit instrument" an election. For Habermas, "Fox's speeches were made

[5] Habermas quotes Edmund Burke on this page: "In free countries, there is often found more real public wisdom and sagacity in shops and manufactories than in the cabinets of princes…."

[6] See Bentham: "The Public Opinion Tribunal is to the Supreme Constitutive what the judiciary is to the Supreme Legislative … Public Opinion may be considered as a system of law, emanating from the body of the people...To the pernicious exercise of the power of government it is the only check; to the beneficial, an indispensable supplement." (Bentham, *Constitutional Code in Harrison 98*)

with the public in mind; "they," the subjects of public opinion, were no longer treated as people whom, like "strangers," one could exclude from deliberations … step by step the absolutism of Parliament had to retreat before their sovereignty…." (Habermas 1989, 66) Here it is critical to note that for Habermas the term public opinion implied that "it was formed in public discussion after the public, through education and information, had been put in a position to arrive at a considered opinion." (Habermas 1989, 66). Habermas later draws from Peel's Tamworth Manifesto (1834) concerning the need for political parties to publish their programs (i.e., platforms) and writes that "Public opinion was formed in the conflict of arguments concerning a substantive issue, not uncritically based on common sense in either naïve or plebiscitarily manipulated assent to or vote about persons. Hence [public opinion] needed a defined issue as its object more than it needed prominent persons" (Habermas 1989, 66-67).

The Formation of Contemporary "Public Opinion"

In no other place was this link between public opinion and public institution more clear, than in the American experiment. Alexis de Tocqueville perceived it in his tour across the continent. And for James Bryce, who wrote *The American Commonwealth* in 1888, the radical equality of all men[7] was turning the United States into what he called "Government by Public Opinion." In ways that would have shaken Madison and the early Federalists, Bryce said: "Towering over Presidents and State governors, over Congress and State legislators, over conventions and the vast machinery of part, public opinion stands out, in the United States, as the great source of power, the master of servants who tremble before it" (Bryce 267 quoted in Fishkin 1995, 73).

The emergence of *opinion publique* from the public sphere had now taken on a life of its own. Bryce saw the need to harness it, but also to be wary of it. In a brilliant psychological reading of modern "opinion formation," Bryce wrote of a hypothetical voter:

> His original impression was faint and perhaps shapeless; its present definiteness and strength are mainly due to what he has heard and read. He has been told what to think, and why to think it. Arguments have been supplied to him from without, and controversy embedded in his mind. Although he supposes his view to be his own, he holds it rather because his acquaintances, his newspapers, his party leaders all hold it. (Bryce 253 quoted in Fishkin 1995, 75).

Yet, despite the suspect and malleable origins of contemporary public

---

[7] The latter designation was all too true until 1920 when the 19th Amendment was passed, granting women to right to vote (See Fishkin 1995, 122-126).

opinion, the democratic process now needed to gauge these opinions. Enter George Gallup and the statistical means to measure public opinion in an almost continuous stream.[8] And whatever the initial democratic hopes of Gallup and his Opinion Poll, it is now a basic fact of political life that such polls serve as constant barometers of "the people's will and interest" in matters as diverse as gay marriage and tax reform.

Moreover, recent social science studies show that people answering Gallup and Gallup-like polls—taking place spontaneously on one's home phone—are prone to produce "pseudo-opinions" that may not even be grounded in reality.[9] Given the growing movement toward "equality" of input in liberal democracies and the extension of this input to direct democratic institution like referenda, this lack of deliberation on the part of the public is cause for concern. Add to this the omnipresence of 24 hour entertainment infused news media, 10 second sound bites, and the large amounts of money being put into special interest advertising, and the problem of public opinion approaches the crisis stage.

Combining Equality with Deliberation: Jim Fishkin's Deliberative Poll®

This crisis is characterized by James Fishkin as a multifaceted dilemma. The problem goes back to the debate between the Federalists and the Anti-federalists in post-revolutionary America. For the Anti-Federalists, American Democracy should be as close to the people as possible. They feared the aristocratic tendencies of a government centered in a capitol far from the everyday concerns of ordinary Pennsylvanians or Virginians. Madison and the authors of the Federalist Papers argued for a strong central government of representatives (Senators) who could deliberate on the common good and avoid the heat of the moment. In broad outline, the Anti-Federalists emphasized a direct democracy in which each citizen had an equal voice in the activities of their government; the Federalists emphasized a representative democracy in which democratically elected members of Congress (and selected members of the Senate) could deliberate as peers on matters of national importance. Behind the desire for deliberation was the Madisonian concern with the tyranny of the Majority (those greater numbers who vote in their own self-interest). The one side emphasized equality; the other deliberation and fear of majority passions. The more you

[8] See "How "Public Opinion Became the Voice of the People" for an engrossing discussion of this phenomenon (Fishkin 1995, Chapter 3).

[9] Fishkin reviews studies by Eugene Harty and Philip Converse in this regard. He also refers to a study in "Public Opinion Quarterly" that sought opinions on the "Public Affairs Act of 1975." Despite the fact that there was no such act, about one-third of the respondents offered opinions. (Fishkin 1995, 80-84).

have equality, the less you have deliberation; the more you have deliberation; the less you have equality. The dilemma can be broken only if you have an institution that manages to combine majoritarian equality with informed deliberation, and does so, moreover, in a consulting rather than plebiscitary way. Recently, James Fishkin has given this approach a concrete and practical instantiation through his concept of a Deliberative Poll.[10]

Deliberative Polling is a process whereby background information on an issue of local or national importance is developed and presented to a scientific random sample of the population. The sample then gathers in small groups, with a moderator, to discuss and deliberate upon the topic. Questions are invariably generated by the discussions and these questions are given to a panel of experts for response. The groups gather again and then answer a list of survey questions. As Fishkin wrote in *Deliberation and Democracy*, "A deliberative opinion poll models what the electorate would think if, hypothetically, it could be immersed in an intensive deliberative process." (Fishkin 1991, 81) In addition to ancillary benefits such as educating and engaging citizens on the issues, the results of these polls can form the basis from which sound policy-making can result.[11]

A central element in the protocols for a deliberative poll is the random sample. Gallup had demonstrated that you need not have very large numbers of people in order to determine the public's views on a matter. A carefully stratified random sample can actually yield a representative sample of the population. This representative sample can in turn satisfy the demands of "equality." The protocol further stipulates that this representative sample (of a City, State, or even Nation) be given background information on the issues to be discussed. Considerable care needs to go into this "issue representation" to ensure that it is fairly presented and that all relevant sides have a voice. Finally, the selected individuals gather in small groups to discuss and deliberate. A moderator guides the flow of the conversation (so that no single voice can dominate, etc.).

Given the random nature of the selection, people naturally encounter others

---

[10] See James Fishkin, *Democracy and Deliberation* (1990) and *The Voice of the People* (1995). The phrase Deliberative Poll is trademarked by James Fishkin and will be assumed throughout.

[11] Fishkin points to the adoption of conservation policies based on a number of deliberative polls sponsored by utility groups in 1996 (See Appendix D in *Voice of the People* p. 220). In this regard, Vincent Price and Peter Neijens raise a number of important methodological concerns in their article "Deliberative Polls: Toward Imporved Measures of "Informed' Public Opinion?" *International Journal of Public Opinion Research* (Summer 1998; 10, 2). While hopeful that these kinds deliberative polls might actually contribute to better public opinion, they argue for more social science research into the area, research that will seek to isolate the various effects that occur in the deliberative process so as to demonstrate the virtues of the process more accurately.

from different socio-economic backgrounds. This causes the group to engage in a respectful dialogue with citizens different from themselves and their group. The result is often a lively exchange of ideas and stories that are refreshing in their variety and sincerity.[12] As questions come up that require more information, the group determines how these questions will be formulated and who will ask them when the time comes to address a panel of experts. The experts are chosen for their background knowledge and represent a range of interpretations of the issues at hand. They are to function as "teachers" rather than "debaters" during the plenary session. The group then returns for one more deliberation before filling out a detailed survey.

"A deliberative opinion poll gives to a microcosm of the entire nation the opportunity for thoughtful interaction and opinion formation that are normally restricted to a small-group democracy," continues Fishkin in *Democracy and Deliberation*. "It provides a statistical model of what the electorate would think if, hypothetically, all voters had the same opportunities that are offered to the sample in the deliberative opinion poll." (Fishkin 1991, 4)

Over the course of the last 17 years, Fishkin has argued that deliberative polling leads to increased objective knowledge of, more engagement in, and better decision-making about social and political issues.[13] Deliberative Polls have been conducted in Britain, Australia, Denmark, and the US, both at the national and local level.[14]

Augmenting Deliberative Polls with Online Tools

In the Fall of 2001 Fishkin gave a presentation at Carnegie Mellon in which he expressed the hope that the Internet could one day support online versions of these polls. The costs in terms of time and expenditures for the National Issues Conversation in 1996 came to 4 days and 4 million dollars. Holding nation-wide deliberative polls on a regular basis was not feasible. And even at the regional level, the additional features of an online version such as asynchronous forums

---

[12] See Iris Young, *Democracy and Inclusion*. Empirical research seems to confirm this (see, for example, Peter Muhlberger "Stealth Democracy, Apathy Rationales, and Deliberation" at communityconnections.Heinz.cmu.edu/papers/abstracts/.html).

[13] Deliberative polls do not usually alter fundamental values (Democrats do not become Republicans, for instance), but often views change on "specific policy attitudes, factual knowledge, and what we have called 'empirical premises' (typically, assumptions about causal connections between policy choices and valued outputs." (*Debating Deliberative Democracy,* p. 30). As John Rawls might say, deliberative polls tend to widen one's reflective equilibrium.

[14] See Stanford's Center for Deliberative Democracy (http://cdd.stanford.edu/).

could add value to even a local face-to-face event.[15] It was from that talk that the seeds for Carnegie Mellon's PICOLA project were sowed.

PICOLA stands for "Public Informed Citizen Online Assembly." It provides a multimedia environment and high-telepresence design for enabling online structured dialogue. At the highest level of Human Computer Interaction, it embeds in its design the notion of "computers as theatre" first described by Brenda Laurel in a book of that name. By redescribing the relation of User to Screen along the lines of Aristotle's *Poetics*, she argues that the user must be brought into the drama of the program and not seen merely as someone outside the screen in need of guidance as to how to navigate the information on the screen. This approach is now apparent in the design of video games when a user can be transformed into a "skier" or "medieval knight." In a similar manner, a user enters PICOLA in such a way as to be transformed into a "citizen" engaged in a community conversation. If we are successful in this, we can make the computer disappear, replaced by a virtual public sphere.[16]

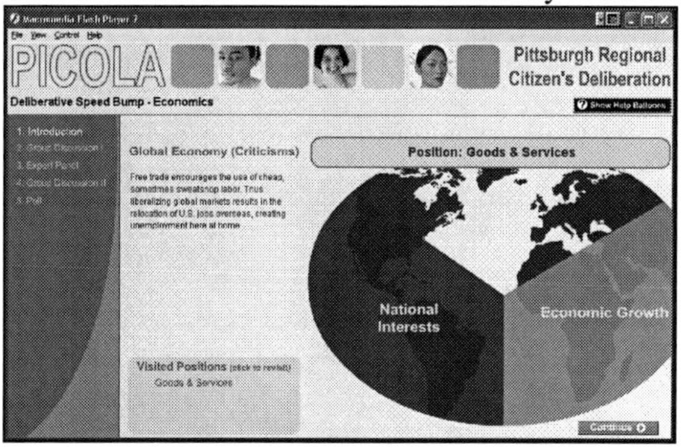

On January 24th, 2004 PICOLA was used by a group of citizens discussing America's Role in the World. This group worked in parallel with a deliberative poll sponsored by PBS's By the People Project (www.pbs.org/newshour/btp/about.html).

After a welcoming and orientation screen, user's register for the program and then are presented with overviews of the day's topics (National Security and Global Economics). We provided what Beth Novick and Benjamin Barber call "deliberative speed bumps" (brief, interactive reviews) prior to the actual online

---

[15] Fishkin's team at Stanford has already conducted encouraging social science experiments with online tools such as Lotus *SameTime*. See "Facilitating Informed Public Opinion: Evidence from Face-to-face and Online Deliberative Polls", Shanto Iyengar, Robert C. Luskin and James S. Fishkin at cdd.stanford.edu/research/index.html. For an overview of online tools for deliberation, go to:
thataway.org/resources/practice/hightech/programs.html.

[16] See Brenda Laurel, *Computers as Theater*. See also Janet Murrary *Hamlet on the Holodeck* as well as Cavalier ("The Poetics of Simulation: An Analysis of Programs in Ethics and Conflict Resolution," forthcoming).

assembly (see their Unchat program at www.unchat.com).

Once people appear at their seats and indicate through a "thumbs up" icon that all is OK, the moderator begins the discussion. An "Ask to Speak" button places the user in a queue and each speaker can have up to two minutes to address the group. The moderator can send text messages to the whole group or simply speak to them as needed. If someone wants to go to the head of the queue, he or she can press a "Jump Ahead" button (but only two such shouts were made available). After the roundtable, we viewed a "netcast" of an expert panel addressing a live audience at the Carnegie Library adjacent to the CMU campus.

Since that demonstration, we have added other features to create a fully integrated online environment for deliberative polling and other forms of structured, democratic dialogue. Currently PICOLA contains registration forms, access to background information, both synchronous and asynchronous multimedia forums, and survey/polling tools. We are exploring designs for a cell phone/PDA interface called Mobile PICOLA (see caae.phil.cmu.edu/picola/).

Fishkin recognizes that these deliberative polls need to have an echo in the media and an influence on relevant stake-holders. The former will allow those citizens not at the event to hear about the discussions and the results of the surveys; the latter requires that, for example, members of City Council (or their staff) attend the plenary session. In the United States, the City of Pittsburgh is beginning to create this new public sphere through the "Southwestern Pennsylvania Program for Deliberative Democracy" (caae.phil.cmu.edu/caae/dp/).

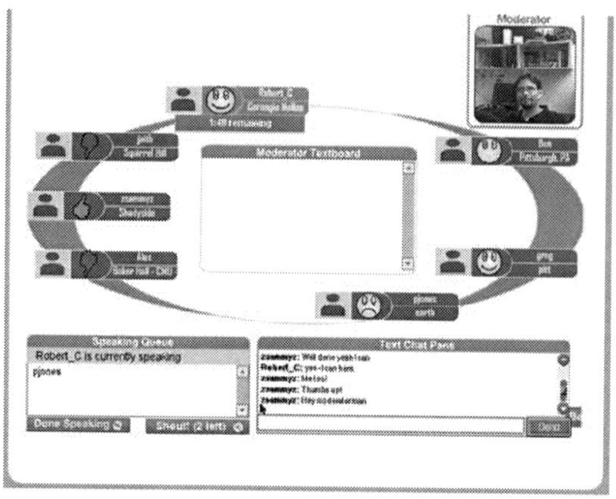

This organisation brings together three Pittsburgh institutions. The first *are the libraries,* which see deliberative polling as an important extension of their civic engagement with the community they serves. Through the development of online versions of deliberative polling, *Carnegie Mellon* sees a natural extension of its interests in the innovative uses of technology. In particular, CMU's Digital Media Lab for Applied Ethics and Political Philosophy and its staff incorporate PICOLA into deliberative polling events and serve as a help center for the use of the technologies developed at the Lab.

Finally, the resources at Public Broadcasting's *WQED Multimedia* and other media outlets will be employed to call attention to both the topics of the poll and the results of the deliberative polling process. It is out belief that universities, libraries and the media, working together with other local organisations and drawing on each other's strengths and resources, will be able to make Southwestern Pennsylvania and the Pittsburgh Region a model for this new form of democratic decision-making—a model that can be tested and shared, eventually, by the nation as a whole and a model that can be incorporated in its own way by other countries, including Thailand, the host site for the 2005 Asia-Pacific Computing and Philosophy Conference.

Through the tripartite approach of face-to-face meetings, augmented by online tools and media broadcast, we believe that the use of deliberative polling has the potential to restructure the public sphere as dramatically as Haberrmas described the transformation of the public sphere during the rise of European bourgeois society. It can return opinion to the status of informed opinion: A public opinion grounded in a *knowledge of the issues* and enriched by a *discussion of the issues*.

References

Amadae, S. M. 2003. *Rationalizing Capitalist Democracy: The Cold War Origins of Rational Choice Liberalism*. Chicago: University of Chicago Press.

Barber, Benjamin R. 1984. *Strong Democracy: Participatory Politics for a New Age*. Berkeley: University of California Press.

Bohman, James, and William Rehg, eds. 1997. *Deliberative Democracy: Essays on Reason and Politics*. Cambridge: MIT Press.

Campbell, James. 1995. *Understanding John Dewey: Nature and Cooperative Intelligence*. Chicago, IL: Open Court.

Downs, Anthony. 1957. *An Economic Theory of Democracy*. New York: Harper.

Fishkin, James S. 1991. *Democracy and Deliberation: New Directions for Democratic Reform*. New Haven: Yale University Press.

—. 1995. *The Voice of the People: Public Opinion and Democracy*. New Haven: Yale University Press.

Fishkin, James S., and Peter Laslett. 2003. *Debating Deliberative Democracy*. Philosophy, Politics and Society; 7. Malden, MA: Blackwell.

Habermas, Jurgen. 1989. *The Structural Transformation of the Public Sphere: An Inquiry into a Category of Bourgeois Society*. Studies in Contemporary German Social Thought. Cambridge, Mass.: MIT Press.

Hamilton, Alexander, et al. 2004. *The Federalist Papers*. New York: Pocket

Books.
Harrison, Ross. 1993. *Democracy*. London; New York: Routledge.
Kymlicka, Will. 2002. *Contemporary Political Philosophy: An Introduction*. 2nd ed. New York: Oxford University Press.
Laurel, Brenda. 1993. *Computers as Theatre*. Reading, MA: Addison-Wesley.
Murray, Janet Horowitz. 1998. *Hamlet on the Holodeck: The Future of Narrative in Cyberspace*. Cambridge, Mass.: MIT Press.
Rawls, John, and Samuel Richard Freeman. 1999. Collected Papers. Cambridge, MA: Harvard University Press.
Rawls, John. 2005. *Political Liberalism*. Columbia Classics in Philosophy. Expanded ed. New York: Columbia University Press.
Riker, William H. 1982. *Liberalism against Populism: A Confrontation between the Theory of Democracy and the Theory of Social Choice*. San Francisco: W.H. Freeman.
Shapiro, Ian. 2003. *The State of Democratic Theory*. Princeton, N.J.: Princeton University Press.
Young, Iris Marion. 2000. *Inclusion and Democracy*. Oxford Political Theory. Oxford; New York: Oxford University Press.

CHAPTER TWO

EXPLORING POLITICAL ONLINE FORUMS IN SAUDI ARABIA THROUGH THEMATIC CONTENT ANALYSIS

YESLAM AL-SAGGAF
SCHOOL OF INFORMATION STUDIES AND CENTRE FOR APPLIED PHILOSOPHY AND PUBLIC ETHICS, CHARLES STURT UNIVERSITY

Background

Saudi Arabia is the custodian of the two holy cities of Mecca and Medina, where Islam emerged, and one of the richest countries in the world in its reserves of oil. These religious and economic factors earned Saudi Arabia a strong position in the Arab world and allowed it to play a leading role in the politics of the region. Saudi Arabia is an absolute monarchy with a population, according to a latest survey, of 22,673,538 including 6,144,236 non-nationals (Al-Watan 2005). The population is predominantly Arab who strictly adheres to Islam.

The 1991 Gulf war affected Saudi Arabia in many ways. The acceptance of the United States" army to liberate Kuwait during Desert Storm upset the Islamic authorities and fundamentalists in the country.[17] To Islamic fundamentalists it was just wrong to use a non-Muslim country against a Muslim country. What made things worse was the government refusal of Bin Laden's offer to protect the country at that time from a possible Iraqi invasion particularly after he came back from Afghanistan victorious in his war against the Soviet Union. The $65 billion US Dollars bill Saudi Arabia paid as its share of the cost of the war caused the Saudi economy to suffer massive losses (Stern 2002). The national budget was reduced by 20% and government employee salaries were slashed some said by about 50%. In addition to making the people in Saudi Arabia uncomfortable in their lives, all of these factors also contributed

[17] The author was in the country at the time.

in some way or another to the sudden emergence of reformists and terrorists in the country.

Saudi Arabia has always been a close ally of United States and the two countries have been good friends since 1940s. The events of September 11, 2001, however, complicated the relationship between them (Stern 2002). The U.S. government, without producing convincing evidence, claimed that fifteen of the nineteen hijackers were Saudis. Some American officials argued that if fifteen hijackers were from Saudi Arabia alone it means the version of Islam that is practised in the country encourages terrorism. Other American officials have gone far to accuse even members of the Royal family themselves of supporting terrorism. As a result of all this, US, indirectly through their media, told the Saudi government, either undertake major religious reforms and revise your educational system, which according to them, lays the ground for extremism or face serious consequences (Butt 2005).

Introduction

Faced by strong pressure from the public, the government of Saudi Arabia in 1999 decided to allow its people access to a censored version of the Internet. The reason for the late introduction of the Internet to the country was the Saudi authority's concerns over the arrival of objectionable material, such as pornography, and sexually explicit material in the hands of the public (Baker 2000).

What Internet users in Saudi Arabia see, however, is not the original page on a server outside the country, but a copy on the computer servers in Riyadh. Saudi Arabia filters all the web traffic that flows to the country by implementing country-level proxy servers. These proxy servers contain a massive database of banned sites (Whitaker 2001).

Today, there are about two million Internet users in the country (Al-Watan 2004) of which about half of these two millions spend their time on the Internet posting or reading messages in web-based forums (JeddahNews.net 2004). Most individuals in these forums post topics to exchange ideas, discuss their public affairs, and analyse political events.

Political online forums (POFs) have become increasingly popular in recent years, particularly after the September 11 attacks on the US and the wars against Afghanistan and Iraq. While freedom of expression in Saudi Arabia is limited and content disseminated from traditional media is censored, the decentralised nature of the POFs and their many-to-many communication nature make them effective in enabling people to express their views and reach others. This article will explore POFs in Saudi Arabia through thematic content analysis.

Online forums and politics in the literature

The relationship between the Internet and politics has been discussed extensively in the literature. In his often cited paper about this topic, Dahlberg (2001) discussed the possibility of a public sphere taking place on the Internet. In doing that he carried out comparisons between people's general practices online and a normative model of the public sphere developed from Habermas' theory of democratic communication. Dahlberg argues that the public sphere that is constituted whenever people enter into political discussions online does not meet all the requirements outlined in Jürgen Habermas' model. This, according to him is because

> First, the increasing commodification of cyberspace threatens the autonomy of public interaction online. Second, reflexivity is often a very minimal part of cyber-deliberations. Third, many online fora experience a lack of respectful listening to others and minimal commitment to working with difference. Fourth, there is difficulty verifying identity claims and information put forward. Fifth, extensive exclusions from online fora result from social inequalities. Finally, discourse tends to be quantitatively and qualitatively dominated by certain individuals and groups.

Despite these shortfalls Dahlberg, however, concludes that the Internet does in fact foster a public sphere because it facilitates rational-critical debates and public opinion formation.

Interestingly, the conclusions drawn from two recent articles by Cammaerts and Audenhove (2003) and Ulrich (2004) coincide with those of Dahlberg. Cammaerts and Audenhove studied four online environments to answer questions related to the usage of the Internet among trans-national social movements from the context of organisation, mediation of participation and public opinion formation. As suggested, they too found that the Internet environments they studied foster public spaces that facilitate online civic engagement despite a number of existing constraints such as those related to access, moderation of interactions and male dominance.

Ulrich (2004), who was also interested in understanding the impact of the Internet on democracy, discussed the various perspectives surrounding the argument about regarding the Internet as a public sphere. While he concludes that the Internet is capable of promoting democratic values, Ulrich also recognises some constraints on the Internet ability. Among the values the Internet is capable of facilitating he listed expression of opinion and generation of dialogue. Dialogue on the Internet, according to him, while not synchronic like face-to-face communication, it is not subjected to the editorial bottleneck that typically characterise the printed media. Dialogue on the Internet also, unlike face-to-face dialogue, provides opportunities for greater thought and comprehensive reflection on the matters discussed.

Methodology

This study builds on an earlier study whose aim was to explore POFs in Saudi Arabia from within the political and cultural contexts of Saudi society. The earlier study was an ethnographic study which took place between 2003 and 2004 (see Al-Saggaf and Weckert 2005). Data for the ethnographic study were collected using silent observation of three political web-based forums used mostly by Saudis. Silent observation was conducted over a period of 14 months from mid-July 2003 to mid-September 2004 and involved recording observational field notes from these forums daily. Data were analysed using a grounded theory approach similar to that detailed by Strauss & Corbin (1998).

The ethnographic study took an interpretivist (or interpretive) approach, which is associated with qualitative research methods. Interpretivists believe that the social world is interpreted or constructed by people and is therefore different from the physical world. They emphasise natural settings and seek to gain deep understanding of the meanings of the actors involved in the social phenomenon under study (Glesne 1999; Sudweeks & Simoff 1999; Williamson 2002).

While the main idea of the present study is to take advantage of some of the features of quantitative content analysis, such as its ability to offer numeric values that will allow reflective inferences of the major themes about a certain topic to be made, the content analysis used in this study also takes an interpretivist approach and is inherently qualitative. Using content analysis in this way (i.e. emerging from an interpretivist framework) seemed to be most appropriate because as in the earlier study, the purpose is also to explore POFs in Saudi Arabia from within the political and cultural contexts of Saudi society and the emphasis is also on understanding the context and meanings that people make of the phenomenon under study.

This approach is also consistent with how content analysis is used in the literature and is often followed in descriptive studies such as the present study (see, for example, Neuman (1994); Aronson (1994); Neuendorf (2002) and Cammaerts and Audenhove (2003)). Conducting content analysis in this way will also allow the themes and patterns related to the studied online forum to emerge from the text and the final result will hopefully be an accurate picture of the characteristics of that forum.

Thematic content analysis

This study relied on thematic content analysis to examine 2000 topics posted to Al[18]-Saha Al-Siyasia (a POF) during September 2004. The unit of analysis is each individual post. While many traditional (i.e. statistical) approaches to content analysis rely on fixed or stable units of analysis, for example, word counts, etc., the focus in this study is on the occurrence of selected themes within each topic. The broader nature and purpose of each topic posted was also taken into consideration during the process of analysis. The decisions of assigning selected themes to topics are made after gaining an idea about the context or contexts each topic fits.

Assigning selected themes to topics based on judgments about their context can no doubt be somewhat subjective. To limit this subjectivity and enhance the reliability and validity of the findings, the researcher triangulated the findings from content analysis with findings obtained from continuously observing the forum for 14 months. (For more information on how this other study was conducted see Al-Saggaf & Weckert, 2005.) In order also to enhance the reliability and validity of the findings, the researcher explained the process of coding and analysis of data in great detail so as to make this study replicable as much as possible (Neuendorf 2002).

Sampling and the studied forum

Al-Saha Al-Siyasia, which is by far the most widely spread POF in Saudi Arabia, receives 700,000 visitors each day (Al Arabiya 2004). On average, each day there appears to be about 70 new topics posted to the forum. While the initial plan was to analyse three months worth of posts designated randomly, unfortunately at the time of collecting data, it was found that Al-Saha Al-Syiasia public archive has kept the posts of only the last two months; September and October 2004. Additionally, at the time of doing the analysis, it was found that the number of topics posted during these two months were huge (4200 topics). As a result, a decision was made to study only one month worth of posts (September 2004). Therefore the sample consisted of all topics posted during that period, which was found to be 2,000 topics. The total number of replies to all of these topics was 15,069. The total of number of times the 2,000 topics were read during the month of September was 6,041,002; with 3,020 as the average number of times a topic was read.

In relation to the issue of ethics in human research, as can be seen from the figures above Al-Saha Al-Siyasia[19] is a hugely populated public space where anyone with an Internet connection can see its content. Recording, analysing and reporting of such public content, where individuals' identities are shielded and no personal information not already available to the public is revealed, is

[18] "Al-" is the Arabic equivalent to "the" in English.
[19] For a detailed description of the design of these forums see Al-Saggaf et al. (2002).

not subject to "Human Subject" constraints as Eysenbach and Till (2001) and Ess & AoIR Ethics Working Committee (2002) argue.

The process of coding

Analysing 2,000 topics is no doubt a difficult task and very time consuming. While most posts were coded with only one category (see Figure 1 below), in a few cases topics were coded with an additional one or two category. In Thematic content analysis it is acceptable that items are code with more than one category at the same time (see, for example, Frankos 2000). This is because the basis for categorisation is the context of the topic, which makes it likely that a topic will belong to more than one category or a theme at the same time. Having said that, however, assigning a topic to more than one category occurred very seldom as in the majority of cases only a single theme was found to be representative of the topic. (This is shown in Table 1 below.)

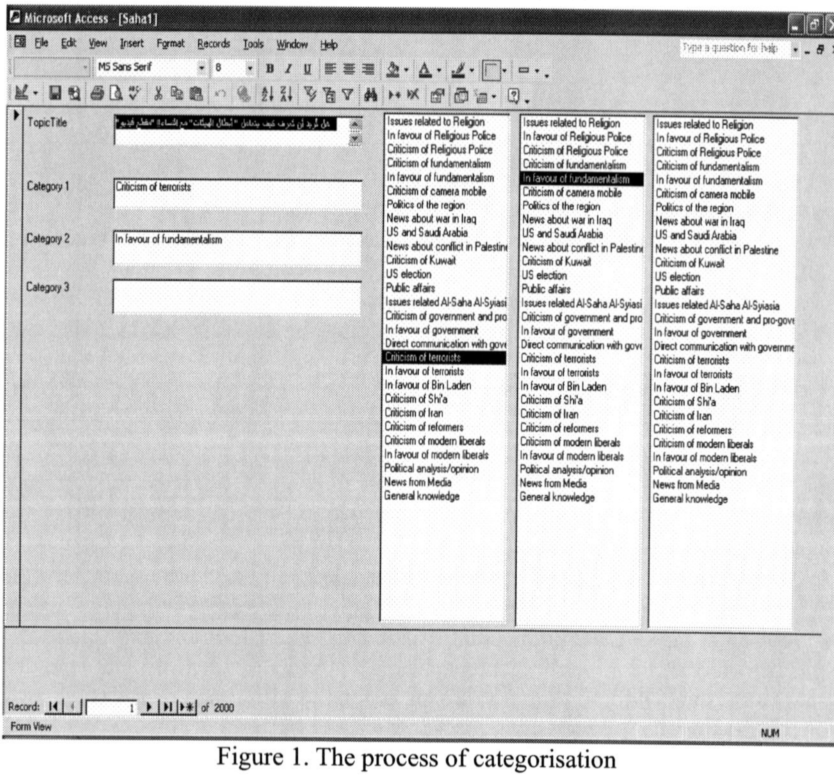

Figure 1. The process of categorisation

First, topic's titles in the form of clickable hyperlinks were copied and pasted into a Microsoft Excel sheet. The clickable hyperlinks titles, which represented links to topics, were copied in groups of 50, as they were located on one page on the forum. Altogether, there were about 40 pages. Next, and after titles were arranged in Excel, they were imported to Microsoft Access as a single table. In Access, three fields were added to that table and named category 1 to 3. The three added categories represented the three possible themes a topic could be assigned. In Access also another table with twenty-seven pre-determined categories was created. (See Figure 1 above for a list of the categories.) The source of the pre-determined categories was the information obtained from the 14 months silent observation of the same forum. It is important to note that additional categories were added as the analysis progressed. This meant that the study relied on both predefined and interactive categories. This introduced a level of coding flexibility as it allowed new and important material that could have significant bearings on the findings to be incorporated into the coding process.

To code data, the topic in Excel was first clicked and read from its location on the Internet. In many cases the replies to a topic were also read to be sure about the context the topic fitted. Second, the topic was assigned a category from a Microsoft Access form in which the researcher created. The form consisted of a textbox that contained the topic title, three empty textboxes to accept the assigning of categories, and three list boxes which had copies of the categories developed earlier (see Figure 1 above). When a topic was selected (after it was read) and upon selecting a category from one of the list boxes, the corresponding empty textbox was automatically filled with the selected category. This process of coding was carried out for all the 2000 topics in the main table.

After coding about 10% of the topics, the researcher stopped coding, reviewed what has been coded and in a separate document explained a number of the categories and made some comments on when to assign them. This process (Neuendorf 2002) provided the researcher with guidelines that helped him in becoming consistent with his decisions about assigning the themes to topics. At the end of this process queries to count the frequency of each category were generated (see Figures 2 and 3 below). Once the coding is done, the researcher examined the findings in light of the earlier study and formed paragraphs that formulated the findings presented in the next few sections.

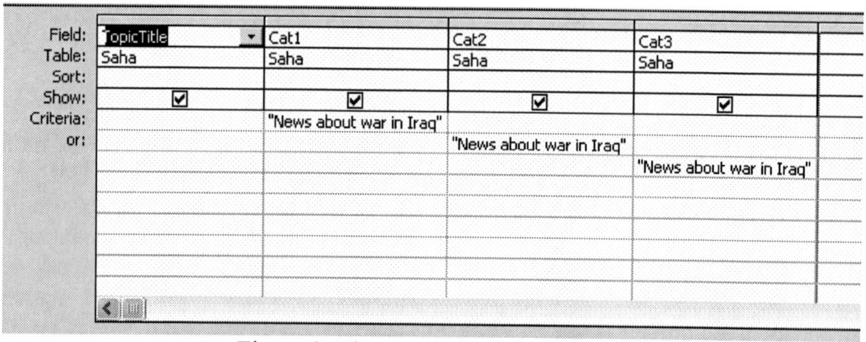

Figure 2. The generation of queries

Figure 3. An example of the result of a query

Findings

The findings of this study are discussed in conjunction with the results of the ethnographic study which was briefly discussed in section 4 above. The interpretations and explanations of the findings below are also made in light of this earlier study. A summary of the findings obtained from thematic content analysis is shown in Table 1 Below:

Category	counts	%
Issues related to Religion	117	5.593
In favour of Religious Police	34	1.625
Criticism of Religious Police	11	0.526
Criticism of fundamentalism	27	1.291
In favour of fundamentalism	125	5.975
Criticism of camera mobile	9	0.430
Politics of the region	101	4.828
News about war in Iraq	243	11.6
US and Saudi Arabia	78	3.728
News about conflict in Palestine	77	3.681
Criticism of Kuwait	13	0.621
US election	5	0.239
Public affairs	379	18.117
Issues related Al-Saha Al-Syiasia	22	1.052
Criticism of government and pro-government	72	3.442
In favour of government	64	3.059
Direct communication with government	41	1.960
Criticism of terrorists	172	8.222
In favour of terrorists	17	0.813
In favour of Bin Laden	23	1.099
Criticism of Shi'a	78	3.728
Criticism of reformers	54	2.581
Criticism of modern liberals	149	7.122
In favour of modern liberals	25	1.195
Political analysis/opinion	65	3.107
News from Media	66	3.155
General knowledge	26	1.243
Total	2092	100.00

Table 1. A summary of the findings obtained from thematic content analysis

Islamic fundamentalism

Nearly 6% of all the topics posted during September 2004 were in favour of fundamentalism against only 1.3% of the topics that criticised this approach to practising the religion. Fundamentalism in the context of this study refers to an approach to practising the religion in which people adhere strictly to the teachings of Islam. The approach is based on a literal interpretation of Islam. People who belong to this category tend to take an uncompromising approach to applying the religion to their lives. The finding that 6% of all forum content was devoted to favouring fundamentalism suggest that people are still generally attached to the way they practise their religion.

Following the events of September 11 and a number of terrorist attacks in Saudi Arabia, this approach came under fierce attack. The source of the attack or rather the openly ongoing criticism was the US (for reasons mentioned in section 1) and modern liberals in Saudi Arabia. From continuously observing Al-Saha Al-Siyasia, it was noted that the reason Islamic fundamentalists were active in the forum is because they felt a need to promote the image of Islamic fundamentalism, subjugate the attack and defend this way of practising the religion. They were also observed to constantly disparage others' ways of practising the religion such as Shia.

It is worth mentioning that the version of the Islam practised in Saudi Arabia is often called "Wahhabism". The "Wahhabi" movement emerged as an effort to wash out Islam of what Abd al-Wahhab, founder of movement, regarded as deviating aspects of modern religious practice. Since its beginning the movement targeted the worship of beings other than Allah it associated with Shiism, and the belief that Allah is everything it associated with Sufi Muslim followers. It banned portraits, gambling, music and dancing.

Shia[20]

About 3.73% of all topics posted to Al-Saha Al-Siyasia criticised Shia Muslims. The figure seems a little large considering Shia Muslims are not welcomed to participate in this forum and their numbers in the Saudi society are small anyway. The Saudi government and indeed the rest of the Gulf States have always feared or hated the Shia Muslims and the country Iran. The war in which Iraq staged and the Gulf States financed from 1980-1988 against Iran is evidence of this fear or hatred.

In Al-Saha Al-Siyasia, Shia was observed to receive continuous criticism from the forum members. In fact, it was observed in the forum that there are members who specialise only in posting topics that condemn Shia Muslims. Some of the topics posted, for example, provided links to movies and images that showed the problems with the Shia ideology and philosophy; others showed pictures of their torture of themselves in memory of the 4th Imam (Ali); yet others showed pictures of their attacks against the Sunni Muslims.

Apparently and since September 11 attacks on US and the "Wahhabi" proponents are on the defensive. One of the reasons for this, in addition to the explicit and severe US criticism of the "Wahhabi" doctrine (Butt 2005), is the

---

[20] "A branch of Islam deriving authority from the prophet's cousin and son-in-law Ali and his appointed successors, the Imams." (*The Penguin English Dictionary* 2003,1288) .

support the Shia and Sufi doctrines received from US and Europe. The reason the Sufi way, which has some links with the Shia approach, is favoured is because the approach itself encourages giving up the material world and dedicating oneself to worship of Allah. This implies that Sufi Muslims generally leave politics and Jihad aside and concentrate only on their worship of Allah, which is what US and Europe want.

Terrorism

While 6% of all the topics analysed favoured fundamentalism, 8.22% of all the topics posted by members during the sampled month were dedicated to criticising terrorists.

These findings are not in conflict with each other because not all versions of fundamentalism encourage terrorism; possibly only one that does. So it quite possible that those who posted topics that favoured fundamentalism also posted topics that condemned terrorism and vice versa.

Terrorism has been a major problem for Saudi Arabia in recent years. Since Al-Kubar attack in 1995 and the country was the stage for more than 50 major attacks. People from "all walks of life" in the forum were observed to hate terrorists for causing the death of many innocent people including their countrymen. Terrorists, according to forum members, took away the sense of security, safety and peace of mind they once enjoyed and that is why they are hated.

It should be noted that there are many participants who sympathise with terrorists in Al-Saha Al-Siyasia. Since terrorists are struggling with establishing their credibility and the grounds for their activities, sympathisers often post topics that aim at improving the image of Bin-Laden and his followers and supporters in front of others. In their topics supporters try to promote the terrorists' propaganda by releasing information about fighters in Iraq and Afghanistan and linking those fighters with them so that people perceive their acts as justifiable as those in Iraq and Afghanistan.

Modern liberals

Modern liberals are another group of people who receive continuous criticism from Al-Saha Al-Siyasia members. Approximately 7.12% of the total forum content in September 2004 was derogatory and belittling of modern liberals. Modern liberals, in the context of this study, are those who adopt a moderate tone in terms of their perceptions of Islam and express this tone in the articles they write in the local news papers and through their appearances in the local media. Modern liberals generally embrace a version of Islam that is less strict in nature and accommodating of others' beliefs. Their aim from what they write or say is, perhaps, to manufacture the consent of people in society so that modernism or secularism is adopted as a way of life.

Al-Saha Al-Siyasia members' criticism of modern liberals makes sense given that many of these forum members are fundamentalists who fear a change to the approach to their life and their religion. This is evident also from the fact that only 1.20% of topics posted were in favour of modern liberals as opposed to 7.12% against them.

Reformers

Reformers are also heavily criticised in the studied forum. While quantitatively only 2.6% of the topics criticised them, qualitatively the criticisms were quite harsh. Unlike in the case of modern liberals in which 1.20% of topics in Al-Saha Al-Siyasia were supportive, reformers, particularly Dr Saad Al-Fagih (the most renowned reformist in Saudi Arabia (Mclaughlin 2003)) hardly had any topics that were supportive.

Dr Al-Fagih, who operates from London, is the director and spokesperson of the Movement for Islamic Reform in Arabia (MIRA) which officially began its operations in 11 March 1996. In the beginning the aim of the movement was to educate the public about, what they believe to be, "prevalent malpractices of the government." MIRA communicated with the public through a high-tech website[21] and a state-of-the-art web-based forum. Today MIRA has judged that it is time to move beyond merely informing the public about their rights and start getting them to demand these rights. One of their biggest achievements so far was getting people out to protest, in 2003 and 2004, in masses in most cities of the country to demand political, economic and social reforms (BBC 2003).[22] People for the first time in the history of this country broke the law that bans demonstrations and marched in most cities of the country to protest against the government handling of the country.

Iraq and Palestine

Around 15.249% of all topics posted to Al-Saha Al-Siyasia during September 2004 were devoted to discussions about the war in Iraq and the conflict in Palestine, with the war in Iraq being discussed more frequently (11.568%). In these topics, forum members posted news about these issues from Satellite television stations, such as Al Arabiya[23], and from websites in which

---

[21] http://www.islahi.net/

[22] The author was in the country when these events took place.

[23] http://www.alarabiya.net/

news were reported by the people who witnessed the events themselves. Members also regularly posted images and video clips of fighting they obtained from non-media sources. In addition, members regularly posted topics to update others about news that were not covered in the media.

In their coverage of the events in Iraq and Palestine members described the events as they were unfolded. Most of the time members also offered commentaries on these events. In these commentaries they synergised ideas and analysed events and frequently provided predictions of what might happen in the future.

Government

While about 3% of the topics posted to the forum were in favour of the Saudi government, about 3.4% were critical of it. This indicates that the government and its supporters are quite active in Al-Saha Al-Siyasia. This could be true even in the case of senior government officials and Royal family members many of whom are said to daily follow this forum. The following excerpt from one of the messages posted to Al-Saha Al-Siyasia shows this:

> This is not strange. What is strange is the minister who does not follow Al-Saha Al-Siyasia. From my point of view, the minister (or the decision maker in general) whose morning program does not include skimming Al-Saha Al-Siyasia and following their links, still lives in the age of King Faisal.

The above message was posted after a member reported that a member of the royal family has responded to a demand discussed in the forum.

The aim of the government officials from participating in Al-Saha Al-Siyasia discussions is to maintain their positive image in front of the public in which anti-government members try to shake. Being present in this forum allows the government to balance things up in terms of clearing away the accusations against them and at the same time promote their political plan, which may include disparaging reformers, terrorists and extremists. Being present in this forum also allows them to keep an eye on what is being said in the forum about them and about the country. Given the vast number of people who operate in this forum, monitoring its content gives the government a chance to gain insight on what is happening in the country.

Public affairs

The findings of this study show that about 18.117% of all the messages sent to the forum during the sampled month were dedicated to discussing people's public affairs. Some members specialised in sharing information about senior government officials whom they accused of corruption. Other members posted topics to complain about things in their country. An example of this was the topic about foreign workers in the country who take jobs from locals causing an increase in unemployment rates. In this sense, the forum is seen as a medium in which people directly communicate with government officials. This is evidenced by the fact that 1.96% of all topics were of this nature. Of course, members use the forum in this way because they know that government officials are present in the forum as mentioned above.

In a number of occasions Al-Saha Al-Siyasia members were observed organising themselves around a political matter that could affect them, and taking a common stand about it in a way not previously possible. An example of this is Crown Prince Abdullah Bin Abdul Aziz's initiative to send Muslim troops to Iraq in an attempt to restore peace in the country (CNN 2004). His plan was heavily criticised by members of the forum from all political currents, who argued that a Muslim nation should not support the enemies of another Muslim nation. According to Al-Quds Al-Arabi (2004), this led the government to "practically" withdraw their initiative. This example clearly illustrates that Al-Saha Al-Siyasia has enabled people to rationally criticise a crucial political decision, reach a consensus regarding it among themselves, and communicate their opposing views about it to government officials. This example also, and others below, suggest that Al-Saha Al-Siyasia may constitute a public sphere which includes some of the elements of the "true" public sphere advocated by Jurgen Habermas (1989) (see a brief discussion about the public sphere in section 3).

Political online forums and the media in Saudi Arabia

As can be seen from the findings of this study, the wide spread use of Al-Saha Al-Siyasia and its popularity in Saudi Arabia, has made it an effective media channel. Even traditional media, such as local television stations and newspapers, have been observed to take some of their news from them. For example, during the capture of Saddam Hussein most international media showed the hole in the ground at the farm where he was hiding. One author in Al-Saha Al-Syiasia, who posted a topic that included that picture, wondered why the colour of dates (brown sweet edible fruit of a palm tree) that appeared in the picture were yellow in a time of the year (December) when the colour of the dates in the trees should look otherwise. His article received a great deal of attention from the readers, as it shed doubts about the truthfulness of the story of the capture of Saddam Hussein. The next day, Al-Watan daily newspaper discussed this article on the front pages.

There is another interesting example that shows the interactions between traditional media and political online forums. The government released, on national television and one of the daily newspapers (AlRiyadh), information about a body of one of the terrorists which they found buried in the desert. In their story, they said that his friends chose not to take him to hospital from fear they would be identified and captured. As a result, they let him die of his wounds. In their story, they also showed the face of the dead person and his body. Obviously, this was an attempt to condemn the acts of terrorists and belittle them in the eyes of their supporters and those who think terrorists acts are legitimate. Terrorists, who operated in Al-Saha Al-Siyasia, were not happy with that coverage and as a result they responded by posting a topic in the forum which caused a sensation. In their topic, they asked why the body of the deceased person did not decompose after two months of burial. Muslims believe that those killed in a battlefield defending their country are martyrs. They also believe that the bodies of martyrs are kept intact in the graves. According to terrorists, if this man's body did not change or decompose after so long, it means he is a martyr. The next day, the government released in Al-Jazirah daily newspaper the medical certificate of the deceased person and some images showing a lot of deformity from decomposition to his body. This interaction between traditional media and the forum clearly shows that the public regards the latter as a valuable media tool. One of the reasons that made information that appear in forums worth reading was the inclusion of images, audio files and video recordings with most of the topics posted to serve as supporting evidence.

Conclusion

POFs in Saudi Arabia are becoming increasingly popular and important in people's lives in Saudi Arabia. Al-Saha Al-Siyasia, which is by far the most widely spread POF in Saudi Arabia, is just one of hundreds of similar forums. One of the reasons these POFs are receiving unprecedented attention in Saudi Arabia is because they serve a specific purpose for people from all backgrounds.

POFs gave average individuals the chance to let their voices heard and get their messages across to others. They offered the public as a whole a medium that is not only widely spread among people and effective in reaching the decision makers in the country, but also not subject to government's control. In addition, reformists found in POFs a channel to communicate with the public and advocate their plans for reform. Terrorists found in them a medium to establish their credibility and the grounds for their actions. The government officials and supporters used them as a medium to maintain their image in front of the public and advocate their plans. Islamic fundamentalists used them as a medium to promote their image and defend their way of practising the religion. Finally, the moderate liberals used POFs to invite people to adopt a less strict version of Islam and adopt secularism as a way of life.

Television, radio and press are, to some extent, in the hands of the government and content broadcasted from these sources has always been geared towards government's plans. The findings of this study show that POFs are the new media which people now have in their hands. Reformists, terrorists, Islamic fundamentalists, moderate liberals and government supporters, are all using POFs to communicate their ideas and plans to others. Their use of POFs is not only making them authors of media content instead of being a passive audience, but also making these tools a valuable source for information about politics and a perfect medium for people to express themselves and influence the political public sphere in their country.

References

Al Arabiya. 2004. *The owner of Al-Saha is in the "Cage of Accusation."* http://www.alarabiya.net/ (accessed August 11, 2004).

Al-Saggaf, Y. & Weckert, J. 2005. Political Online Communities (POCs) in Saudi Arabia. In S. Marshall, W. Taylor & X. Yu, eds. *Encyclopaedia of Developing Regional Communities with ICT.* USA: Idea Group Reference.

Al-Saggaf, Y., Williamson, K. & Weckert, J. 2002. Online Communities in Saudi Arabia: An Ethnographic Study. *Proceedings of The Thirteenth Australasian Conference on Information Systems ACIS*, 4-6 December, Melbourne, Australia.

Al-Quds Al-Arabi. 2004. *Saudi Arabia withdraws its initiative practically.* http://www.alquds.co.uk/ (accessed August 4, 2004).

Al-Watan. 2005. *The main features of the population in Saudi Arabia.* http://www.alwatan.com.sa (accessed February 25, 2005).

—. 2004. *To improve services for subscribes and promote research.* http://www.alwatan.com.sa (accessed April 15, 2004).

Aronson, J. 1994. A Pragmatic View of Thematic Analysis. *The Qualitative Report* 2 (1). http://www.nova.edu/ssss/QR/BackIssues/QR2-1/aronson.html (accessed April 15, 2004).

BBC Arabic. 2003. *Arrests in Riyadh after demonstrations demanding reform.* http://news.bbc.co.uk/hi/arabic/middle_east_news/newsid_3192000/3192170.stm (accessed April 1, 2005).

Butt, G. 2005. *Saudi Arabia: Political overview.* http://news.bbc.co.uk/1/hi/world/middle_east/3784879.stm (accessed April 1, 2005).

Baker, C. 2000. Culture Shock: Saudi Arabia. *The Internet Society.* http://www.suite101.com/article.cfm/5785/4021 (accessed June 30, 2000).

Cammaerts, B., & Audenhove, L.V. 2003. *ICT-Usage among Transnational*

Social Movements in the Networked Society: To Organise, to Mediate & to Influence. ASCoR. Amsterdam: Amsterdam Free University.

CNN. 2004. *Powell welcomes Saudi plan on Iraq*. http://www.cnn.com/2004/WORLD/meast/07/29/powell (accessed July 29, 2004).

Dahlberg, Lincoln. 2001. Computer-Mediated Communication and the Public Sphere: Acritical Analysis. *JCMC, 7* (1). http://www.ascusc.org/jcmc/vol7/issue1/dahlberg.html (accessed September 7, 2004).

Ess, C. & AoIR ethics working committee. 2002. *Ethical decision-making and Internet research: Recommendations from the aoir ethics committee*. http://www.aoir.org/reports/ethics.pdf (accessed April 3, 2003).

Eysenbach, G. & Till, J. E. 2001. Ethical issues in qualitative research on Internet communities. *British Medical Journal* 323 (7321): 103-1105.

Frankos, C. E. 2000. *Interactivity and Community: Social Interaction in An Online Community*. Unpublished MA thesis, California State University: UMI.

Glesne, C. 1999. *Becoming Qualitative Researcher: An Introduction*. 2nd ed. New York: Longman.

Habermas, J. 1989. The Public Sphere. In *Jurgen Habermas on Society and Politics: A Reader,* ed. S. Seidman, 231-236). Boston: Beacon Press.

JeddahNews.net. 2004. *Where Do You Spend Most of Your Time Over the Internet?* http://www.jeddahnews.net (accessed September 7, 2004).

Mclaughlin, S. 2003. The use of the Internet for political action by non-state dissident actors in the Middle East. *First Monday*, 8 (11), November 2003. http://www.firstmonday.dk/issues/issue8_11/ (accessed March 30, 2005).

Neuendorf, K. A. 2002. *The content Analysis Guidebook*. Thousand Oaks: Sage.

Neuman, L. 1994. *Social Research Methods: Qualitative and Quantitative Approaches*. Massachusetts: Simon and Schuster.

Stern, M. 2002. *Threat of U.S. action against Iraq further tests key anti-terror all*. http://www.signonsandiego.com/news/nation/terror/dispatches/20021222-9999_saudimain.html (accessed March 28, 2005).

Strauss, A. & Corbin, J. 1998. *Basics of Qualitative Research: Techniques and Procedures for Developing Grounded Theory*, 2nd ed. Thousand Oaks: Sage.

Sudweeks, F. & Simoff, S. J. 1999. Complementary explorative data analysis: the reconciliation of quantitative and qualitative principles. In *Doing Internet Research: Critical Issues and Methods for Examining the Net,* ed. S. G. Jones. Thousand Oaks: Sage.

The Penguin English Dictionary. 2003. 2nd ed. London: Penguin Books

Ulrich, J. 2004. The Internet as a political public sphere. *Proceedings of the ICT and learning in regions conference*, 1-2 June 2004, Aalborg Congress Centre, Denmark.

Williamson, K. 2002. *Research Methods for Students and Professionals: Information Management and Systems.* 2nd ed. CSU, Wagga Wagga: Centre for Information Studies.

Whitaker, B. 2001. Losing the Saudi cyberwar, *The Guardian*, 26 February.

CHAPTER THREE

TECHNOLOGY AND JOURNALISM IN THE MARKET: EXPOSING PRESS CORRUPTION BY COMPARISON

AARON QUINN
UNIVERSITY OF CALIFORNIA

In the West journalists typically believe they have more press freedom than anywhere else in the world—particularly in the United States because of its Constitution's First Amendment, which legally protects freedom of speech. Consequently, Western news audiences believe they have the best—or at least most uninhibited—news available. This paper will argue that although the United States and several other Western countries have substantial legal protections from Government interference, the modern institutional trend of media concentration is by default narrowing the scope and perspective of news coverage, sometimes deliberately via corporate edict and other times unintentionally because of the ever-shrinking number of independent news sources. This creates an institutional moral problem because Western news audiences believe they receive genuinely free speech—that is, reportage free of substantial ulterior agendas whether they are corporate, political or otherwise. What they actually receive is reportage moulded by corporate interests and unofficial government censorship allowed by a flawed institutional scheme that panders to corporate growth. Thus, in some cases, media concentration is not only morally questionable, but responsible for outright corruption via its malignant institutional structure.

Freedom of the press—especially press freedom that is garnered through independent ownership—is central to enabling the fulfilment of journalistic moral role requirements and thus is a crucial part of generating a balanced democracy. When journalism fails to fulfil this goal, one can assume journalism's institutional role-moral responsibility is at least neglected if not corrupt. The mistaken assumption Western news audiences have rests in their broad trust of their favourite news organisations as stewards of the public rather than, as it is in some cases, self-interested businesses. Fortunately, the public availability of the World Wide Web can play a crucial role in uncovering Western journalistic self-censorship, and, as we shall see, also provides a hint at what is at least a possible remedy for it.

This remedy will focus on the relevance of truth to journalism, particularly a method of mining truthful or at least relevant information in an institutional scheme involving a *constrained* competitive marketplace. This notion derives from well-known "marketplace" theories introduced by John Stuart Mill and John Milton. Implementing a *constrained* marketplace approach in journalism, I will argue, is necessary to both complement contemporary legalistic press protections threatened by self-censorship, and to overcome the truth-suppressing propaganda resulting in part from media concentration. This argument is best supported by observations journalism scholars have made in the world's broadest marketplace of ideas—the World Wide Web—through which those who are connected can see ideas, both bad and good, and evaluate them in order to make decisions central to public life.

This examination will comprise of the following components in this order: First, a brief history of information technologies that have liberated information for a broader public and for a greater good. Second, we will review Herman & Chomsky's (1988) and Bagdikian's (1997) arguments outlining how media concentration has both decreased the number of ideas in the marketplace, and, just as importantly, has created a corrupt, propaganda machine that in some cases supports partisan political and corporate interests. Third, we will review cases that expose media self-censorship and propaganda and examinations of deleterious Western journalistic trends that become clearer when compared with non-Western news practises by using the Web as a resource. Last, we will review how a marketplace scheme exposes, and, to a limited extent, may reduce Western self-censorship if it is constrained by particular rules.

Historical Roots of Techno-liberation

For centuries information technology has played a central role in promoting public empowerment through information. An early example of liberating information technology emerged when power was wrested from the Roman Catholic Church after the printing press was made widely available. It eliminated the church's monopoly on Biblical interpretation and distribution of religious ideas. As individualised interpretations of the Bible arrived along with wider distribution, so came the Protestant Reformation (Irish Independent 2004), which spilt Catholic detractors into various Protestant faiths.

Several centuries later, radio technology was used to transmit news about people and events happening behind the "Iron Curtain;" the project was called

Radio Free Europe (Ruckstuhl 2003). It allowed many who lived under communist rule to understand domestic news without the hindrance of state-controlled censorship (Although it was arguably filled with Western propaganda instead). In the 21st Century, the names and places have changed but the classic problem remains: The free press is threatened by elitist propaganda and journalistic self-censorship—that is, the press barons often choose not to fully inform the public while instead creating news agendas that fraudulently benefit political and corporate allies (McCombs & Shaw 1972). Motivation for such corruption should be clear in the subsequent discussion regarding corporatisation of news media.

Telecommunications technology (including the Web and satellite TV) is essential for examining the differences among news providers, but its availability is fairly new and rapidly increasing its reach (e.g., with satellite news subscriptions and increasing Web use). Since Western news agencies are the de facto source of news for Western audiences, these audiences are rarely aware that non-Western news sources such as *Al-Jazirah* may fulfil their "public-interest" duties better than traditional Western agencies—particularly with international-issues coverage. Therefore, we will examine to what extent Western news agencies are respecting the *public-interest rights*—covering politics, safety issues, business, etc—of their audiences. But first we will set the stage for this analysis by reviewing some of the deleterious business practises that in large part have led to the problem of self-censorship.

News Censorship as a Corporate Agenda

Both Herman & Chomsky (1988) and Bagdikian (1997) make strong claims about corporate control over social and political agendas via the mass media. They point out a clear conflict of interest, which exists in any industry where a fundamental news media role is subverted by a corporate profiteering agenda with self-regarding economic and political goals. Although one may rightfully claim journalism is hitched with an unavoidable conflict of interests because of its internally conflicting values as both a business and a news provider, there is substantial opportunity for discretion by journalistic organisations in terms of balancing their economic and editorial priorities. Thus journalists use the metaphor of the "unfailing wall" between editorial (news makers) and advertisers (revenue producers).

Clearly the moral imperative in journalism favours whatever priority furthers the goal of providing the audience with information that is crucial to the public's interest. As a broad but sometimes bending rule, fulfilling essential newsgathering duties always trumps economic considerations unless the economic need is so foundational that it threatens the existence of the news agency.

Practically speaking, this means that news agencies are not morally justified in cutting newsgathering staff to lower operating costs if it marginalizes the quality of news coverage merely for the sake of *excessive* profit (For example, American news agencies averaged a 22 percent profit margin in 2004, up 4 percent from the previous year, yet still managed to cut one percent of its overall news gathering staff). This rule would not stand, for example, if a journalist were compelled to overlook a damning investigative story because it would damage the reputation of one of its advertisers, thus lose its advertising revenue, even if it were, by chance, likely to go bankrupt the news agency as a result. In other words, there are certain newsgathering principles that no consequentialist ethic can override.

This is one facet of an institutional role morality, which, ideally, requires an institution to fulfil its socially important institutional purpose before concerning itself with fiscal surpluses. However, as we will see, it is not always direct economics—such as refusing to expose an advertiser of wrongdoing—that dictate the agenda of corrupt news media, because political preferences, for example, may claim an equal or greater censorship consideration by media owners if it is to their business advantage.

Motivations for and examples of censorship based on Political Preference

Motivation for news media to give political preference is difficult to prove, but there are certain logical motivations for news agencies to have a preference, including, for example, a political administration's stance on media ownership restrictions. George Bush's appointee to the U.S. media regulator, the Federal Communications Commission, loosened media ownership restrictions from a single corporate entity being limited to reaching 35 percent of the population to a 45 percent share, which is in line with the Bush administration's pro-corporate economic etiquette. In Australia, the Howard government has pushed three time for media ownership changes, first in 1997 when Howard proposed that all ownership restrictions be abolished, which he did again prior to the election in 2001, and yet again in 2004 when he took a more moderate approach by suggesting a "modernisation" of media ownership that, like the Bush administration, allows giant media corporations to at least expand (Murphy & Lewis 2004).

As media ownership exists in Australia today, limitations for owners are hardly strict. In terms of single-medium ownership, Australia allows the following: Television station owners may control stations that reach up to 75 percent of the *entire* Australian population. Radio ownership is limited to no more than two radio licenses per license area—a license area is, for example, the metropolitan area of Melbourne (Broadcasting Services Act 1992). Newspaper ownership is only limited in cross-ownership regulations. Cross ownership regulations in Australia are as followed: One entity may not own both a television and radio station in the same license area; nor can one entity own a television station and a newspaper in the same license area; nor can one entity own a radio station and newspaper in one license area. Some examples of current Australian media concentration are as follows: The Ten Group, Limited, owns television interests that can reach up to 65 percent of the Australian population. News Limited, chaired by Rupert Murdoch, controls 68 percent of Australia's capital city and national newspaper market, among other substantial television and radio interests.

Still regulators insist media ownership laws effectively protect free speech. "The major effect of the laws is to prevent the common ownership of newspapers, television and radio broadcasting licences that serve the same region. The purpose of the legislation is to encourage diversity in the ownership of the most influential forms of the commercial media: the daily press and free-to-air television and radio" (Jackson 2003, 1). Despite these already vast concentrations of media ownership, there is still a bitter feud over loosening existing restrictions. "This debate has been fuelled by the impact of new media technologies, a number of inquiries proposing regulatory changes, and the self-interest of those media organisations that report the controversy. The [Howard] Government has long indicated that it believed the rules to be anachronistic, and its policy… contained a commitment to amend cross-media and foreign ownership restrictions" (Jackson 2003, 1).

U.S. media ownership debates nearly mirror the Australian dispute. Although population share percentage allowances in the United States are lower—for example, the maximum television share per owner in the United States is 45 percent up from 35 percent in early 2000—there is a trend to increase that share and loosen cross-media control. For example, it is now legal to own a newspaper and television station in a market that has more at than four television stations. In markets with more than nine television stations, there is no limit on cross-ownership.

Local television ownership rules have changed so that companies can own three stations in markets with 18 or more television stations. In markets with five or more television stations, a company can own two stations (FCC 2004). Regardless of the specific numbers, there certainly seems to be a tacit agreement between media companies and certain political administrations. It is our task to examine several unethical political preferences displayed by news agencies in order to demonstrate how this preference marginalizes the essential journalistic value of free speech and its correspondence with truth.

Big Business, Big Bias

Now that we have established logical motivations for politically sensitive media ownership, we can filter through substantial cases in which political sympathy has surfaced in recent years. Fox News Corp., owned by Australian Rupert Murdoch, was chronicled in an Australian Broadcasting Corporation documentary as giving direct orders to its journalists to sympathise with President Bush in his 2004 re-election campaign. Also, Sinclair Broadcasting Group, based in the United States, broadcast a shamefully biased documentary depicting John Kerry as a wartime failure during his tenure as a Vietnam soldier. The film was broadcast within days of the 2004 presidential election with the alleged intention to discredit Kerry's reputation as a war hero.

Fox

The 2004 documentary, *Outfoxed*, featured testimony from ex-Fox employees who claimed there were directives from Fox management to deliver positive news about the U.S. war in Iraq, especially in light of the failed effort to find Weapons of Mass Destruction. David Korb, a former freelance writer for Fox, claimed that when the public image of U.S. troops in Iraq became bleak, a senior Fox producer gave instructions to the writers not to mention the number of dead troops or anything about troops being under enemy fire. There were also details in the documentary of written directives from news managers demanding particular story angles and biases on sensitive war-related stories.

Sinclair Broadcasting Group

Just weeks before the 2004 U.S. Presidential election, the Sinclair Broadcasting Group, which controls 62 television stations in the United States, announced its plan to air a prime time news feature in which presidential candidate Sen. John Kerry would be ridiculed for allegedly renewing practises of torture against prisoners of war in Vietnam in the early 1970s. The filmmaker proposes that Kerry's post-war comments criticising American troops for abusing innocent Vietnamese sparked a renewed interest in torture against American prisoners of war in Vietnam in the early 1970s. Kerry supporters claim the film is a propaganda tool filled with false claims and loose causal relations to the violence; namely, that Kerry's anti-war stance actually directly incited torture.

Notwithstanding the validity of claim, the timing of the broadcast of *Stolen Honor*—slated to be anywhere from two weeks to four days before the presidential election depending on location—was potentially disastrous for

Kerry in light of the closeness of the race according to pre-election polls (virtually a dead-heat as of 17 October 2004). Sinclair's decision was so contentious, in fact, that it fired its Washington D.C. bureau chief after he publicly voiced his dissent over the decision to air the film so close to the November election. Jon Lieberman, the sacked journalist, called the film "biased political propaganda, with clear intentions to sway the election" (Folfenflik 2004, 1).

Since Sinclair released its plan to air the program, it drew the ire of the Kerry campaign, media watchdogs, pro-Democrat groups, and a litany of other objectors, some of whom have lodged lawsuits over "equal-time" laws required for any campaign messages broadcast within a certain proximity of a U.S. presidential election. It was also noted that Sinclair is one of President Bush's most substantial campaign donors (Anonymous, *Sydney Morning Herald* 2004).

New Journalism, Bad Journalism or Corruption?

At best, Fox and Sinclair made errors in news judgment. At worst, they developed intentional tools for political manipulation. Some evidence points to the latter. For example, a prime moral obligation of any news agency is to deliver information in the public's interest in a timely manner. If Sinclair found *Stolen Honor* a formidable, credible piece of news, it neglected a moral imperative to broadcast the piece soon after it was first completed. Therefore, one must question Sinclair's decision to suppress politically crucial information for weeks if not months. Moreover, how is one to assume there is not political motivation for airing the program just days before an election?

Not even a Sinclair bureau chief could believe that, so he risked and eventually lost his job to reveal Sinclair's wrongdoing. During the firestorm that developed after the programming announcement, filmmaker Michael Moore even offered Sinclair a free copy of his documentary, *Fahrenheit 911*, as a counter-balancing biased broadcast, which Sinclair promptly dismissed because, Sinclair claimed, it is too critical of President Bush.

However, if Sinclair did consciously decide to make a biased political decision, it not only violated the standard journalistic values of impartiality and balance, but, by doing so, also acted corruptly because of the deceitful, covert, and damaging way in which the partisan feature was packaged as news. Because of the severe consequences of news and partisan politics overlapping into a single package, it is necessary to make the following distinction about corruption in reference Sinclair and Fox's aforementioned organisational decisions.

First, a corrupt act is an intentional act of deception by a corruptor to the corrupted for the purpose of self-regarding gain. Second, this deceptive act must break a fiduciary duty of trust between the corruptor and the corrupted. Third, an act must also undermine a legitimate process or role to such an extent that trust in that process or role is significantly damaged. This usually implies that the act must be habitual—not just one-off—however, if the consequences of a one-off action are so severe or so well publicised that it significantly damages a trust-contingent relationship, this may also be considered corrupt.

First let us review the intentional act of deception: Sinclair and Fox, by airing a politically-charged program falsely labeled as fair and balanced journalism, would be intentionally deceiving their audiences for the purpose of self-regarding gain; namely, for making an effort to bolster their political allies and reap any of its perceived rewards.

Both Fox and Sinclair also fulfill the second condition for corruption: breaking its fiduciary duty of trust to the public. By marginalizing its truth telling duties to further its business or political interests, both companies gave the public reason to lose faith in them and to some extent in the institution of journalism.

Both companies also meet the third condition for corruption. Sinclair is a 62-station conglomerate whose broadcast can reach up to 25 percent of the entire TV audience in the United States. As a result, it would not only influence hordes of potential voters, but it would dissolve the legitimacy of Sinclair as an organisation, and, more broadly, would injure journalism's institutional reputation by undermining journalism's foremost institutional process: gathering truthful information. Fox, which is much larger than Sinclair, and is an international news provider, would have a broader reach yet, thus even a larger negative impact on its viewers as well as the subsequent loss of trust. Moreover, if the *Outfoxed* documentary was correct, Fox made a practice of violating the standard journalistic values of impartial and balanced reporting; hence, although Fox only needs satisfy one of the two possible tiers of the third condition for corruption, it not only satisfies the more rare condition of corruption on grounds that it made a massive one-off impact in undermining a legitimate institutional role, but it also satisfied the standard condition of habitually violating standards which are part of a legitimate institutional role. Therefore, both companies have the potential to fulfil each condition for corruption not only because they are habitual offenders of institutional values, but they are also offensive on a grand scale because of their enormous market shares.

The Importance of the Corruption Distinction

So why bother with claiming corruption? Broad exposition of unethical actions is crucial for professional criticism in that it identifies and exposes behaviour deleterious to people, organisations and institutions in its own unique way. However, limiting this criticism to the rubric of honesty, integrity, or professionalism is a disservice to differences in severity among various possible moral measurements. Thus, different types of unethical behaviour require different categories since there are clear distinctions in both characterising wrongdoing and compelling reform. Often, corrupt actions fall under the category of gravely unethical behaviour and thus ought to receive the level of attention and reprimand a gross ethical violation deserves. Moreover, types of unethical behaviour—irrespective of severity—have different components and ought to be classified separately for the sake of conceptual clarity. And for this reason, we develop taxonomies for ethical violations, which include corruption (Spence, Miller & Roberts 2005; Thompson 1994). This allows us to better understand unethical actions" effects and to determine culpability or prevention, among other goals ethicists have.

In the following section, we will continue with our analysis of corrupt censorship, but in a slightly different context. The following cases will employ comparative practises between Western and non-Western news agencies, which will show how current Western practises may be less free than either Western journalists or Western audiences generally believe, and certainly less free than their functional and moral role requires.

Comparative News

It is when diverse news sources are compared that we can find valuable information in their differences. Since the Web is a valuable resource for comparing many things, including news—thousands of reputable, diverse news sources are available—it offers us a sample of how a more diverse journalistic marketplace might look. Essentially, no single national interest can control all news media available on the Web. With that in mind, many scholars and news viewers have made cross-cultural news comparisons that show differences in journalistic content, style and context—some of which shows passive or deliberate self-censorship. For example, Fox's omission of troop deaths from many of its news reports where it would have been both relevant and contextually necessary.

One example that illustrates the divide between Western censorship and non-Western resistance to censorship came after U.S. Secretary of State Colin Powell attempted to use his influence to censor *Al-Jazirah*, a Qatari news agency, after it repeatedly aired an exclusive interview with Osama bin Laden after the Sept. 11 attacks. Powell claimed that any news agency that aired bin Laden's messages must by default oppose the American war on terrorism. "The request from Powell followed a formal diplomatic protest by the U.S. Embassy in Qatar, which expressed concern about the station's alleged anti-American bias and its repeated airing of a 1998 exclusive interview with Osama bin Laden" (Simon 2002, 64). Moreover, swarms of American journalists initially decried *Al-Jazirah*'s decision to air the bin Laden interview, whereas after the fury passed, most major networks chose to air the tapes despite their initial misgivings.

A number of press coverage trends support the claim of a U.S. government-sponsored "chilling effect" on American media. Moreover, some speculate the unusually strict informal censorship in the United States set a precedent for increased censorship by governments abroad. For example, Simon (2002) suggests U.S. censorship influenced a moratorium on photographing the president of Uganda; a ban on anti-American reporting in China; beatings of journalists covering anti-U.S. demonstrations in Indonesia; arrests of journalists in Benin for reporting that bin Laden had contacts there; and a banning of Israeli state radio for airing live interviews with Palestinian militants (Simon 2002, 64-65). Although there is no direct evidence of a causal relationship between U.S. censorship and censorship abroad, there is at least a plausible view that it may exist.

Conversely, and despite the fact that it is state-funded by Qatar's premier, *Al-Jarirah* has bucked a number of international censorship trends that are typically thought by Westerners to be more pervasive in the Arab world. While U.S. journalists and U.S. allies may in some cases censor Anti-American sentiments in news, *Al-Jarirah* has on numerous occasions delivered news that could be perceived as both anti-American, pro-American, or nation-neutral news (Hanley 2004). Moreover, it has been both applauded and criticised for its descriptive prose and explicit photography that graphically portrays the brutality of war. While Western news—especially in Australia and the United States—rarely shows images of wounded or dead U.S., Australian or British soldiers, but it blatantly flaunts deaths of the opposition. *Al-Jarirah* regularly pictures casualties and fatalities on all sides giving a more accurate account of wartime violence.

As we have seen, there are often substantial differences in style and content between typical Western news agencies and non-Western news sources such as *Al-Jazirah*. Despite *Al-Jazirah*'s lack of official free speech protection, it appears to resist the growing Western trend to pander to political interests. And it is through emerging technology—and that technology's broadening of the marketplace—that we can make this comparison and reveal some of Western news's flaws. But before we institute a marketplace theory, we must understand the limits of it. On the one hand, we want a system that allows as many *legitimate* market participants as possible. On the other hand, this system must be restricted so that the increase in participants is geared towards participants that are both somewhat independent (not conglomerate) and qualified (actual journalists).

The Marketplace, Capitalism and Democracy

Democratic processes have a number of functions that are aimed at creating and maintaining a just and fair society. Free speech, including freedom of the press, is an ideal component of a democratic society as it creates a forum for critical evaluation of issues that matter and people with power, as well as a place to exchange whatever transient hodgepodge of information which happens to be vital to the public's interest. In this section, we will discuss how best to maximise the amount of truthful information in the public sphere.

The Marketplace of Ideas

We will begin this discussion of democracy by introducing the Marketplace of Ideas—a thought experiment that culminates with independent citizens making choices about public issues among a theoretically open and competing group of ideas. Chronologically speaking, John Milton (1644) first formulated the Marketplace for the purposes of press liberty—he addressed England's Parliament to persuade it to support unlicensed printing, thus reducing the centralisation and censorship of the printed word. He further stated that all ideas are welcome in this marketplace, even bad ideas that seem to have little or no merit in themselves. Despite an idea's potential lack of inherent value, he said, there is something to be garnered by it.

"Bad meats will scarce breed good nourishment in the healthiest concoction; but herein the difference is of bad books, that they do a discreet and judicious reader serve in many respects to discover, to confute, to forewarn, and to illustrate. Whereof what better witness can ye expect I should produce, then one of your own now sitting in Parliament, the chief of learned men reputed in this land, Mr. Selden, whose volume of natural and national laws proves, not only by great authorities brought together, but by exquisite reasons and theorems almost mathematically demonstrative, that all opinions, yea errors, known read, and collated, are of main service and assistance toward the speedy attainment of what is truest" (Milton 1644, 44).

Milton explained that bad ideas become more obvious and make good ideas more evident when they are juxtaposed alongside one another as if they were fruits in a marketplace. Moreover an open marketplace of ideas prevents infringement of independent voices (and prevents any loss if ideas to the public) in the case that a new view may be suppressed by censorship before it can be publicly evaluated for its merit. It is through examining as many ideas as possible that we are likely to find that "what is truest" (Milton 1644, 44). John

Stuart Mill (1859) also had much to say on the notion of an independent voice in the marketplace, illustrated in *On Liberty*. "If all mankind minus one were of one opinion, and only one person were of the contrary opinion, mankind would be no more justified in silencing that one person, than he, if he had the power, would be justified silencing mankind" (Mill 1859, 79).

What Mill addressed in terms of individual voice—and Milton for book publishing—is obviously valuable to journalism. Since Western journalism's business practises have become increasingly concentric in the last several decades, and because the number of independent sources and perspectives available to the public are now so few, there is a very limited market from which to choose. However, Web and satellite technologies now allow publics abroad a more expansive marketplace, minimally constrained by Western economic trends, political priorities or standard journalistic practices. Moreover, because the public interest of Americans, for example, overlaps with many of the interests of other nations, there is a limited but substantial shared public interest, which allows people with Web or satellite access to cross-examine news from diverse sources. With all of the choices offered through telecommunications passages, we have a glimpse of a marketplace we should strive for (with some modifications) outside of the virtual world.

Ideal modifications to a standard marketplace include at least two broad constraints relevant to ownership concentration and journalistic quality. First, legislation must be implemented to drastically reduce media concentration in many Western nations, including Australia and the United States. It is virtually impossible in the current market for an upstart news publication to compete with media conglomerates in the same way it is nearly impossible for a penniless presidential candidate to defeat a competitor who has tens of millions of dollars at his disposal. Although there is technically a market into which an upstart company can attempt to compete on idealistic grounds, it cannot compete on economic grounds, thus most upstarts are suffocated by the conglomerates' overwhelming resources.

Second, we must constrain the marketplace to publications that are sufficiently journalistic for the sake of ensuring at least a minimum of quality. Rogue bloggers and journal publishers (Web or otherwise) may well have valuable messages for the public, but they are not journalists until they have the requisite minimal education, training and experience, thus they ought to seek non-journalistic methods for the diffusion of their messages (clearly the modern Web presents a challenge to this limitation).

Essentially, this means traditional journalistic publications such as newspapers ought to be specially grouped for sale, say in journalistic kiosks, exclusively for the sake of identity and ease of public understanding. Non-journalistic publications, in this scheme, must be sold in a separate kiosk/rack/section by news vendors. For broadcast journalism, there ought to be both a required number of hours devoted to news on free to air channels daily some of which ought to be prime time. Essentially, requiring certain journalistic standards and separating journalism from non-journalism is meant to counteract the potential publishing anarchy to which a more independent press could be vulnerable. That is, although journalism ought to be open to independent agencies, it ought not to change its institutional ideals.

With this kind of marketplace, truth or truths of some kind are more attainable—or so that is what the marketplace is meant to deliver. Hence, as truthtelling is the seminal journalistic value, this marketplace is a morally justified mechanism insofar as it best provides the cross-sections of information necessary for audience members to make informed decisions about public life.

The idea of putting moral constraints on a competitive market—which in a way limits free speech—is not without philosophical precedent. Moral Philosopher and Economist Adam Smith proposed that corporate goals inherently lead toward monopoly thus corporate speech and market freedom does not deserve absolute protection because it does not advance the overall wellbeing of society (Kerr 2000). Thus, using Smith's criterion, it appears morally acceptable if not morally imperative to impose market constraints as long as they promote overall social wellbeing, because, in fact, these constraints broaden the competitive nature of the market.

However, corporations typically counter-claim that conglomerated corporate success increases social wellbeing more than diverse markets. Economist Milton Friedman (1962), for example, would likely insist journalism's primary responsibility would be to its companies' shareholders, not its audience. "...[T]here is one and only one social responsibility of business—to use its resources and engage in activities designed to increase its profits so long as it stays within the rules of the game, which is to say, engages in open and free competition, without deception or fraud" (Friedman 1962, 133). What Friedman fails to acknowledge here is the moral validity of the rules of the game, which clearly disregards the many negative consequences imposed by large corporations to non-shareholders. However, the primary reason moral consideration must reach beyond the wellbeing of shareholders is a teleological one; that is, journalism's primary function is to inform the public, not to maximise profits.

Audience Responsibility

Thus far we have established several key features to support this paper's argument for considering news media decentralisation. First, concentrated news media contribute to trends of either deliberate or indirect political bias resulting in a reduction of free speech. Second, we have seen that the Web offers a forum through which we can identify censorship trends. Third, the Web also offers an imperfect but valuable example of how a broad marketplace is superior to a narrow marketplace in terms of finding diverse news content and perspectives. Fourth, a marketplace constrained by strict ownership concentration limits and quality control might increase the public interest value of news. However, this evaluation must take a further step to show how this constrained marketplace can be effective in a Western democratic market.

A key point of discussion here is the concept of *citizen*, and whether he or she is willing to use information in a way that will help her make decisions about public life. For example, one could imagine a hypothetical polis in which the journalistic institution performed its role ideally, but remained irrelevant because of an apathetic citizenry that felt neither obliged to debate public issues, nor even read, watch or listen to reports about them. Ultimately, unless citizens consider and then (in)act on information accordingly, the *co-operative goals* of the journalist-citizen relationship go unfulfilled.

This raises several possible journalist-citizen relationships, assuming, for argument's sake, that journalists are sufficiently informing the public. First, is the struggling relationship that seems to be prominent in much of the West today—a somewhat apathetic citizenry that has minimal interest in hard news; several possible causes for this disinterest have been postulated, including the public's distrust for news media, a greater interest in gossip news, a disconnect with public affairs or a combination thereof.

The second scenario, and a more ideal relationship, requires news media to inform the citizenry, which empowers it to make thoughtful, independent decisions based on this information; broadly construed, this is an *informed democracy*. Third, is the notion that the press informs the citizenry, which then results in public deliberation, and, until this deliberation occurs, a satisfactory "truth" is not likely to be reached—this describes a *deliberative democracy.* Critical Theorist Jürgen Habermas (1989) wrote extensively in the late 20th Century on deliberative democracy by emphasising the significance of the *public sphere.*

Habermas described an ideal *public sphere* as consisting of private individuals who congregate in public places to form a public body. This public body is responsible for educating itself and debating on matters of public importance, thus eventually "publicising" political issues the way political scholars back to Immanuel Kant envisioned. This provides—at least to some extent—a modicum of political transparency and then a public voice in response to the political choices of the governance (1989). It is this public voice that attaches normative value to public decisions via the natural filtering of social norms and values through news consumption and deliberation.

But clearly there are practical limits to this thesis, as even Habermas admits, including the issue of whom the public sphere consists, how large a portion of the population it consists of (Who does not participate in the public sphere and why?), and to what effect this publicity leads (Does it actually contribute to public policy?). Habermas likened the public sphere to a coffeehouse gathering in which all members of the public are welcome, those who choose to come invest time in reading newspapers, news magazines and other journalistic media, and willingly engage in debate about the issues presented in those media (Cowan 2004). Therefore, the ideal Habermasian public is both informed and deliberative. Although it would seem the deliberative democracy is preferable, it is historically often too intellectually cumbersome to popularise for long periods. However, as far as the practical stance we must take here, journalists must only concern themselves with making an informed or deliberative democracy *possible*, not with making it function, as they have neither the purpose nor power to establish this. However, what the institution of journalism is capable of is organising its structure and function in such a way that diverse content and perspectives are available to the public so that it is possible for it to be informed and to deliberate effectively if it chooses.

However, in the world in which we live, journalists are not consistently fulfilling their theoretical public-regarding duties sufficiently, much less ideally, which leaves us with the necessity of making a practical teleological (goal-oriented) evaluation. For example, journalists must decide whether to provide the public with what it apparently wants in terms of ease of use—in many cases entertainment or superficial news briefs—or to give the public what it needs to fulfil its ideal democratic potential—information that provides sufficient illumination on matters of the public interest. News media are tempted to deliver entertainment and lite news because it is profitable. On a practical note, however, the alternative—informing or promoting deliberation (public journalism)—has historically been profitable as well (although not nearly as profitable as entertainment), and more adequately fulfils the ideal role of journalism—informing the public on matters of the public interest. In the following section, we will see how an alternative view of democracy based in a limited "marketplace" might provide the best option.

Conclusion

It is essential to classify the moral issue of politically-motivated self-censorship as an institutional and/or organisational problem. Too often institutional and organisational moral issues in journalism and other professions function with impunity because ethical analysis is geared toward the conduct of individuals. This allows for corrupt practises to flourish because there is difficulty in establishing fault and accountability when institutional or organisation problems arise. Thus the emphasis in this paper on identifying and uncovering corruption on the part of journalism's concentric institutional structure (near monopolies) and its organisation participation (Fox, Sinclair, etc).

Fortunately for the public, the Web can serve as a dynamic and liberating forum that has probably not yet revealed even a scintilla of its functional potential. Its marketplace capability is just one of a seemingly infinite number of advantages, but one that is limited both in structure and in effect. Even this imperfect marketplace, unfortunately, will only last as long as its corporate masters see fit to publish for free or for a low cost. For now, thousands of credible, sophisticated news sources are accessible without charge, but that is expected to change. And with that change will go many a person's capability or motivation to access the marketplace. Although a free-of-charge media system is not necessary for a marketplace to thrive, the current no-cost system demonstrates what informational advantages are possible in a competitive forum where independent, divergent views are readily available. Right now, the Web may be the only place where this occurs.

Ultimately, one must hope for a regulatory structure, as Adam Smith might have proposed, that would limit corporate conglomerations that undermine the public's interests, specifically its interest in finding truth about matters central to public life. For now, however, the Web is offering a glimpse of what could be in a futuristic regulated competitive market system.

References

Bagdikian, B. 1997. *The Media Monopoly.* 5th ed. Boston: Beacon Press.

Broadcasting Services Act. 1992. http://scaleplus.law.gov.au/html/pasteact/0/136/top.htm. (accessed November 25, 2004).

Carter, B. Oct. 21, 2004. Broadcaster's stock picks up after change on Kerry film. *New York Times,* sec. A, p. 27.

CNN, anonymous. 2004, April 30. McCain rebukes Sinclair "Nightline" decision. *CNN online.* http://edition.cnn.com/2004/SHOWBIZ/TV/04/29/abc.nightline/. (accessed October 21, 2004).

Cowan, B. 2004. Mr. Spectator and the Coffeehouse Public Sphere. *Eighteenth-Century Studies* 37.3: 345.

Federal Communications Commission. 2004. Media Ownership Policy Re-Examination. http://www.fcc.gov/ownership/. (accessed December 6, 2004).

Folfenflick, D. Sinclair fires D.C. chief who spoke out. *The Baltimore Sun.*

www.baltimoresun.com (accessed October 20, 2004).

Friedman, M. 1962. *Capitalism and Freedom.* Chicago: University of Chicago Press.

Habermas, Jürgen. 1989. *Structural Transformation of the Public Sphere,* ed. Thomas Berger. Cambridge: MIT Press.

Hanley, D. 2004. Al JarirahWorld Forum takes a hard look at Freedom of the Press. *Washington Report on Middle Eastern Affairs* 23. 8: 14-17.

Herman, E.S. & Chomsky, N. 1988. Propaganda Mill. *The Progressive* 52. 6: 14-18.

Herman, E.S. & Chomsky, N. 1988. *Manufacturing Consent.* Toronto: Random House.

Irish Independent Newspaper, editorial. Sept. 15, 2004. via *Factiva* database, Sept. 22, 2004.

Jackson, K. 2003. Media Ownership Regulation in Australia. *Parliament of Australia, Parliamentary Library e-brief.* http://www.aph.gov.au/library/intguide/SP/media_regulations.htm (accessed December 5, 2004).

Kerr, R. 2002. Impartial Spectator in the Marketplace of Ideas. *Journalism and Mass Communication Quarterly* 79. 2: 394-416.

McCombs, M. & Shaw, D.L. 1972. The agenda-setting function of mass media. *Public Opinion Quarterly* 36: 176-187.

Mill, J.S. 1859. *On Liberty.* New York: Dutton.

Miller, Roberts & Spence. 2005. *Corruption and Anti-Corruption.* Upper Saddle River, New Jersey: Prentice Hall.

Milton, J. [1644] 1932. *Areopagitica.* Ed. J. Hales. Oxford: Clarendon Press.

Murphy, K. and S. Lewis. 2004. Media Laws Deal Possible. *The Australian,* Oct. 13, 2004.

Ruckstuhl, C. E. 2003. The beginning of electronic warfare. *The World & I* 18.10.

Simon, J. 2002. Look who's inspiring global censorship. *Columbia Journalism Review* 40.5: 64-65.

Sydney Morning Herald, *Anonymous.* 2004. Anti-Kerry film sparks row. www.smh.com.au (accessed October 20, 2004).

Thompson, D.F. 1995. *Ethics in Congress: From Individual to Institutional Corruption.* Washington D.C., Brookings Institution.

CHAPTER FOUR

A GLANCE INTO THE LIFE OF A COMPUTERIZED GENERATION: A CASE STUDY ON THAI TEENAGERS LIVING IN BANGKOK

CHULANEE THIANTHAI
DEPARTMENT OF ANTHROPOLOGY AND SOCIOLOGY, CHULALONGKORN UNIVERSITY

Introduction

From abacuses to self-portable eyeglasses, then computers and now the development of artificial intelligence robots, humans have used their imaginations to improve their creations, to learn from them, as well as be affected by them. Throughout time all of these inventions have been perceived to be an inseparable part of human evolution. We are moving towards the era of computerisation. Along with being able to read, write, and calculate, computer literacy has become one of the most important skills that people of this generation need to have (Komonmas 1991).

Computerization began in Thailand in 1964 when the first two computers were brought into the country to be used in the National Statistical Office and the Chulalongkorn University (Tiotuyco and Pakorn 1985; Komonmas 1991; Malaiwong et. al.1998). Ten years later the demand for imported computers in Thailand had increased in both the public and private sectors at an annual rate up of to 75% (Tiotuyco and Pakorn, 1985). Brands such as Universal Automatic Computer (UNIVAC, now known as IBM), Apple, NCR, CDC, and Compact, were the first groups of computer sellers that had made their way into the Thai market (Komonmas 1991).

At present, computer usage in Thailand has come to the fourth wave. In Srichai's article written in 1991, he describes the evolution of computer usage in Thailand and divided it into four periods of time. In the first wave, computers in Thailand were being used for routine statistical entry work. Then computers were used for management of information systems in private organizations in the second wave. They helped many managers analyse data, made important company decisions, and predicted future trends. In the third wave, computers had improved its capability in terms of becoming much faster, being more friendly to the user, and most of all, equipped with accessories (e.g. diskette) that could transfer what was being saved onto another computer. This third wave was the "decision support system age." The fourth and present wave of computers in Thailand has now entered the era of utilizing computers as a way to communicate and get connected to the world. A recent study conducted by NECTEC in 2004 predicted that by the year 2005 over 6 million Thai people will use the internet (The Internet Index of Thailand [online] 2004). Moreover, many high-technology related computing options such as electronic mailing system (e-mail), tele-conference, and special purpose computers have been created to serve people in all occupations and age groups. These types of waves might occur similarly to other countries, but there are also variations that could be found.

Many researches have shown usage of computer-related variations can occur in many forms: differences of computer usage occur in each gender, class, age, and level of education, and can even be found between developed and developing countries (NECTEC 1999; Piyanut 1999; Puntawee 1999; Sirintorn 2002; Information Technology Report Column 2003; Ornuma 2002). While the developed countries use the internet mainly for communicating, the developing countries are likely to use their internet for information purposes. Thailand seems to fall in between the developed and developing country's category; that is they use the internet as a means for communicating through e-mail as well as searching for information (Piyanat 1999). In Thailand, teenagers ranging in age from 15-19 were voted the age group who uses the internet the most (The Survey of Information Technology 2003). There is also a significant difference in how each age group perceives and uses their computers in their everyday life (www.thaiparents.net 2002). For example, children use computers to play games, teenagers use them to get connected to the world and to make new friends in chat rooms, adults use them for work and search for data, and the elderly use them indirectly as health equipment (Sarawut 2003; Monchai 2004; Theppipit 1990; Pentip 1997; Srisab 1999). Likewise, there is also a difference between generations, those who were born before the computerization age and those in the computerized generation, that is while the former tends to use their internet for task-oriented goals, the latter age group uses the internet to entertain themselves or serve their interests (Paratasillapin 2001).

But what is it really like to grow up in the computerized generation? There are a few studies and books written based on what has happened in Western society (Tapscott 1998; Anderson and Tracey 2001; Williams 2002). Rarely have there been studies conducted among the computerized generation who live in a transitional society in Asia. I aim to use this research to fill this gap. This research will explore on the unique characteristics of this new generation of Thai teenagers that grew up in the computerization era. In addition, this research also aims to provide the readers with insight into what computers mean to them, what their daily internet-related activities are, and how they foresee themselves living in the future of advanced computer technology.

Setting, Population, and Research Methodology

This qualitative research was conducted in Bangkok, known as the city of angels. Bangkok, the capital of Thailand, is where the social lives of the city people, business flow, important political institutions, and high technology advancement is located. In order to study the lives of a computerized generation, Bangkok is a perfect setting for this research objective. The intensive data collection was gathered during October-December of 2004. Sixty male and sixty female middle to upper class high school students, who were born in the years 1985-1989 when computers started becoming an essential part of urban living, were invited to join in this study. These teenagers came from four different public high schools in central Bangkok.

Ethnographic research techniques, namely, participant observation, interview questionnaires, in-depth interview, pictorial ethnography, and secondary data research helped conceptualize some of the important aspects of the computerized generation. First, I observed different types of computer and internet techno-related activities Thai teenagers engaged in. Secondary research techniques such as reading former studies, Thai newspapers, books, and computer related articles also helped in understanding how computers have had an impact on both this age group as well as Thai society. Information gained from both of these research techniques was used to form the structured question of an interview questionnaire.

One hundred and twenty interview questionnaires were handed out to students in accordance to the purposive and snowball sampling techniques. The interview questionnaires were composed of three sections. The first section consisted of demographic questions, such as age, year born, gender, level of parents' income, level of education, level of parents' education, and number and ownership of a computer at home. The raw data was coded and entered in the SPSS program to show the demographic characteristics of the sample population. The second section of the questionnaire covered open-ended questions aimed at answering each of the research objective questions. Pictorial ethnography technique was combined with the third section of the interview questionnaire. The third section of the questionnaire was a blank page for students to draw pictures and describe how they foresee their future living with computer and high-technology. After the picture was drawn, I would ask the students to explain the pictures in their own words, presenting their imaginations through verbal conversation through exploring and devouring on the drawings they have drawn.

I selected thirty students who filled in interesting information in the questionnaires as well as showed interest and willingness, to participate in an in-depth interview. I used interviews to dig deep into these interesting answers and clarify some of the information that was unclear. In-depth interviews helped provide rich information of how the students perceive their generation to be different from other generations.

I then selected eight people out of thirty students who participated in the in-depth interviews as specific case studies. The participant observation technique was being used again in following the lives of these eight students. I aimed at being involved in learning the computer-related activities of these students, such as chatting finding friends in chat rooms and pursuing their interests on their favorite websites.

Table 1. The number of participants participating in each research methodology

Research Techniques	Number of Participants
Interview questionnaires	120 participants
In-depth interviews	30 participants
Intensive participant observation case studies	8 participants
Total	**120 participants***

Notice: * Teenagers who participated in in-depth interview and represent intensive participant observation case studies were the same group of teenagers who participated in the initial interview questionnaire.

To keep the student's personal data confidential, all names mentioned in this research are replaced by pseudonym.

Discussions

Being One of the Computerized Generation

When asked to join in this study, there are no doubts or second thoughts from these Thai teenagers that they may not be a part of the computerized generation. Many have asked what is the net generation? How is the computerized generation different from the generation of their parents, like the baby boomers? I think that the answer to this question is much clearer, well explained, and amusing coming from these teenagers' own words.

Researcher: Please describe and define the children of the computerized generation?

Teenagers: The children who grew up in the computerized generation are…

- Those who prefer to use computers for entertainment more than to go out, run around, and play with friends outside the house.
- Those who live a convenient life and use technology heavily in their everyday lives.
- Those who can think, develop their thoughts, keep up with new technology, and are able to use technology in a beneficial and moral way.
- Those who live and interact with others (human) in both real society and through the internet community.
- Those who think that it is necessary to use computers in their daily lives for doing things for themselves including using it for relaxation.
- Those who were born and grew up in a society that uses high technology. Everything that they do from the time they were born till the time they die involve technology and computer control.
- The generation who can utilize computers the most.
- The new generation have no boundaries of their thoughts, freedom in expressing oneself, and able to form infinite numbers of networks through cyberspace.
- Those who use computers a lot and cannot live without it.
- Those who are modern and advanced people who can keep up with the world at just one touch of the computer in their home.
- Those who have computers and know what internet means.

Above were selected answers that came from Thai teenagers most of which represent these common themes. Although these teenagers might not be familiar with the phrase "computerized generation" that refer to their generation image, they have a pretty good idea of what their generation is. Not only a contrast between generation's thoughts, feelings, abilities, and activities occur, but also some of the descriptions make us think about how in the real world there are also a digital gap between those teenagers who have these technology and those who have not.

Daily Life Activities of the Computerized Generation

Living in the computerized generation means more than just being surrounded by high technologies and computer gadgets. To these students, it means that technology and they are inseparable in their everyday activities. Since they can remember, their parents have used technology to raise them, they learned to use it, and accepted it as a part of their routine. Just like learning to brush their teeth everyday in the morning. It is so normal and natural. This is because technology has always been there for their generation. Therefore, it is not surprising that during the interviews, it was hard for these teenagers to name their activities associated with technology and computer-related gadgets. However, when given time to think of their daily schedule they have come to realize the enormous involvement between themselves and their reliability on these high tech instruments.

Table 2. Illustrates two of the examples of the computerized generation daily lives on the weekday (left) and weekend (right).

Monday-Friday

Time	Technology and Computer-related Activities
6.20-7 am.	Waking up because of digital alarm clock
	Using electric tooth brush
	Taking a water heated shower
7.00-7.40 am.	Driving to school or taking BTS then bus to school
	Viewing electronic billboards on BTS platform or new the time reduction traffic lights
	Listening to music on the car radio or using portable MP3 walkman radio
8.10-11.50 am.	Using lift to the classroom
	Turning on lights, air conditioning, and sound system/microphone in classroom
	Drinking water off the drinking fountain
11.50-1 pm.	Eating school meals or Western food sold at school usually cooked by electronic cooking utensils
	Calling friends from cell phone/sending messages
	Playing games or use computer chat program
1- 4 pm.	Similar to 8.10-11.50 am. activities
4-8 pm.	Using cell phone to call or receive calls from parents/ friends
	Driving in a car, bus, BTS back home
	Turning on air conditioning at home
	Eating dinner with family where most of the electric kitchen utensils are used

8-10.30 pm.	Watching/listening to channel V or Mtv channels and doing other things
	Taking a shower/ using razor
	Turning on their computers to chat, play games, search/print information for school report
10.30 pm.	Sleep

Saturday-Sunday

Time	Technology and Computer-related Activities
7-9 am.	Taking a water heated shower
	Reheating the food in microwave
	Turning on the television
9.30-11.00 am.	Taking the BTS to tutorial classes
	Learning tutorial lessons through the videotape
11am.-2 pm.	Eating lunch and buying fashionable fruit-blended drinks at Siam Square
	Maybe going to the movie theater with friends
3-5 pm.	Taking BTS home
	Turning on air conditioning and calling friends
	Watching television or playing DVD , Listening to the radio or playing CD
	Surfing internet, checking e-mail
5-7 pm.	Doing homework with the help of computer in searching for data
7-9 pm.	Driving out for a family dinner
9-10 pm.	Taking a shower and blow drying hair
10-11.30 pm.	Surfing internet browsing through favorite websites
	Using internet to chat with friends
11.30 pm.	Sleep

The daily life activities of the computerized generation were not much different between the weekdays and weekends. The type of technology and computer-related activities are slightly different according to socioeconomic levels of each household and each gender group. Teenagers whose parents have less income may live life with few types of convenient technologies than those whose parents have higher incomes. While some girls like to chat and send messages through (MSN) internet to friends, most boys prefer to play online games. Overall, this group of teenagers often spends 2-4 hours per day using computer. Every teenager from this study at least has one computer at home, only three did not. However, they are planning to buy one soon. Computer is found in every room in their household, in their bedroom, their parent's room, living room, guestroom, their family study room, even some reported that their parents left the old one in the junk room of there house.

These Thai teenagers felt that they are blessed to be born with these technologies. In Thai society, we often throw around using word like "information technology" every now and then. The Thai sense of what technology means to this generation has extended to mean more than just things that were invented by human beings. Technology to them has transformed into various shapes of creations that are made from fundamental applied science knowledge, which are put to use to benefit human living conditions. The word "technology" also equates to things that are new, different, modern, useful innovations, being developed from the old version, show progress through time, facilitate and make things much more faster and easier, and serve human needs. Technologies that most fascinate these teenagers are computers, clonning, nanotechnology, walking and talking robots, cell phones that can take pictures and connect to the internet, Blue tooth technology, modern space shuttles, the invention of internet, MSN, new Sony walkman CD player, iPod, new movie stimulation, including James Bond's detective gadgets.

Almost all of the students in this study think that adults use computers differently from their generation. For example, adults often use computer as a type of working machine especially used for typing and making presentations, whereas teenagers use computers for multi-purposes, especially for personal entertainment, searching information, communicating with friends, and to open up their mind to the outside world. Since computer usage by adults is more work or business orientated, adults seem to use computers to earn more money whereas teenagers time in using computers often relate to spending (time and sometimes money). Most adults usually spend quality time and less time on their computer only when it is necessary; whereas, teenagers used them for a long time everyday. Many adults view computers as a knowledgeable source of information, but sometimes these teenagers view computers as a boring time killer for nonsense usage. Moreover, teenagers think that adults are fixated on the old computer programs they are familiar with and are reluctant to learn to use new programs unlike themselves.

These Thai teenagers seem to be born with special talents in making the most out of computer usage and use it for multi-purposes. The types of activities include communicating through e-mail, chat, MSN, ICQ, typing, calculating, searching for information for reports, submitting homework to teachers, working on a part-time data entry job, studying online, playing online games, making new designs, finding pictures for art projects or presentations, make personal diary and picture online album, listening or loading music/ring tones, read online newspapers, looking up new fashion trends, check movie schedules, reading website that has cartoons and transcribed Korean and Chinese novels, keeping up with entertainment news and celebrity gossips, making new friends, dating, voting and expressing their opinions on societal issues on webboards, selling things, consulting problems on anonymous online services, watch movies or television clips, download new programs, and build their own websites.

The Symbolic Meanings of Computer and the Feelings of Being in Cyberspace

> Computer is my partner both at work and at play
> —Prim, a 15 years old, very talkative female student

> Computer is a part of my life, my heart and my soul
> —Kang, a 17 years old male high school student

To these Thai teenagers, computers are more than just a machine. Computers are their technological friends, which help entertain and relax them, get their jobs done, relieve stress, and especially cure their boredom and loneliness. Not to mention cater to their everyday needs, computers help makes their lives more convenient. When playing with computers it made them feel that they are a part of the future generation that is modern, self-independent, clever, smart, intelligent, and in the same group as the in-trend people. Unlike other advanced technologies, the interaction they have with computers is much easier and more user friendly. Like playing with your extraordinary friends, they can feel the excitement, rush, and happiness.

Their sense of self in cyberspace also provides them with unique feelings. Although cyberspace is such an abstract concept when it comes to explaining and imagining its being in the universe, Thai teenagers can feel the existence of it as if they are in the other world. In this study, only a couple of teenagers stated that cyberspace to them is just another space in time where they did not feel anything special. The majority of the teenagers have described their experiences in cyberspace as extraordinary:

> It is my private world where I can expose my real being and freedom. I can do anything I want. Nobody knows who I am. It is like being in another world
> —Tuta, a 15 year old, a female high school student

> I feel good, it is fun…it is really a fun place to meet new friends.
> —Off, a shy 17 years old, tenth grade male student

It is fun, like being in a very large space with infinite boundary.
—Wit, a 19 years old boy

I can feel the excitement of seeing new things in cyberspace.
—Pam, a twelfth grade high school student

So fun to be in it that it makes me forget time.
—Oat, a 16 year old girl who likes to chat on the internet

As we can see cyberspace is a new spacescape made for today's generation. Teenagers like it because they can explore themselves, attached to boundless rules, and it is unpredictable to what they will find in this world. The feeling of being in cyberspace and the symbolic meanings of computers are closely related to one another. Computers serve as a media that transfers these teenagers to another world and transform their imagination into visual substance. In this other world, the world of cyberspace, they can enjoy themselves, make their own rules, and have control over it. Learning how these teenagers perceive computers and give meanings to their cyberspace is among the first steps in providing a deeper understanding of the computerized generation.

Characteristics and Concepts Thai Teenagers Use to Identify as Computer-Related Experiences of Their Generation

People of each generation tend to have certain shared characteristics. The usage of computers and internet has affected the ways this generation think and act. Through participant observation, data gained from interview questionnaires and in-depth interviews, I have noted several computer and internet influences on the young Thai generation, which are distinct from other generations as seen below.

Like interactive technology and like having control over things

Teenagers stated that it is more fun to use interactive technology. This is a huge part of why they love computers so much because most of the computer related activities are interactive such as chatting, sending messages back and forth with their friends, and playing online games. Technology that provides a one-way interaction like radio or television seems to have decreased in popularity when being compared to computers. Moreover, this generation likes to be in charge and take control over technology. They like to be active and be the planner, not the passive follower.

Create their own private world

Although the teenage years is the age when one likes to have one's own privacy, but for this new generation they want even more. Using a lot of computers has kept them from outside activities with friends. Today's teenagers are satisfied with creating a one-on-one relationship with a computer. They create their own world by interacting with their computers.

Self exploration

They learn more about themselves, what they like and what they don'tlike through the internet. This is a huge part of the self-exploration process. The internet allows them to find more information and make them become more knowledgeable in the topics that most interest them. Thus, it is much easier to learn who you really are through the help of the internet.

Freedom in emotional and intellectual self expression

The cyberworld allows these teenagers to share and express their deep emotional feelings and open up intellectually on web boards. To some extent, they carry back home these behaviours of such freedom in expressing and defending one's mind that made adults view their generation as being rude.

Investigative and innovative

Search engines have become a teenager's best friend. They know that anyone can post things on the web, so they do not believe or just stick to one website for information. Instead, they use search engines to find out about other sources and opinions. This investigation gives them a clearer picture and encourages them to form their own opinions based on the information that they have uncovered on their own. On the other hand, the innovative ideas also are simulated by search function in the internet. The ways that teenagers learn new things have triggered their creativity imagination and make it into reality.

View the world broadly with an open-mind

During the interviews, many teenagers have mentioned how their generation is shaped to be more open-mind because they see more and learn more about people's differences. Cyberspace has opened their worldview and it has also broadened their mind in ways that are seen as the generation that is more open than others when it comes to thinking out of the box or accepting differences.

Get connected with their friends, with others, with the world

More and more teenage activities prove that they like to "get connected" to the others. Calling friends on cell phone for no reason have become a casual behaviour. They also want to know what is going on in their society including what is happening around the world. Being connected gives these teenagers the sense of being a part of the society they live in. No matter if that society is as small as their close circle of friends or as large as being a part of the members in world community.

Have new sense of timing

Technology has shaped them to experience new sense of time. We are moving into the era that internet might not be enough, but hi-speed internet is what is the most desirable! Teenagers reported that the fast response of technology have made them realized how time is one of the most valuable things in life. Waiting and having patience in getting what they want have become shorter than their parent's generation. This is because they are used to the convenience advanced technology provides for them. If loading a new game only takes them seconds, how can they patiently wait to receive a letter through regular mail?

Enjoying multiple identities

When they go online, these teenagers get a chance to be someone they might not be, including not showing their real identity to others. It is fun just to make up a fake name and use it in a chat room. It is like playing a game where you can be whoever you want to be and try to fool the people on the other side of the screen. A girl can simply go online and discuss heavy topics like politics or fake to be a man and court other women. However, some teenagers also reported that they still think sometimes doing these types of things can have negative consequences and they do not enjoy it as it is unmorally wrong. In sum, the Thai teenagers enjoy the new experience of having multiple identities to a certain extent.

Use computers more for leisure than for work

Teenagers and kids in the computerized generation often use their computers more for leisure purpose than for work. They use computers to reduce stress and entertain themselves. Some may wonder what stress these teenagers have. Most of the teenagers in this study reported their stress often came from school and doing lots of homework.

Have high sense of alertness

This generation of teenagers, whether they like it or not, are automatically alert to news and happenings. This is because news now travels faster in this era, teenagers like to be a part of what they hear, especially news of others' well-being and wars. This sense of alertness also includes following the trend of when the new types of high technology will be out. You can often hear question like "when will the new Nokia cell phone addition be out?" or "what does the new model look like?" spreading around in teenagers' conversation.

Want options

The way technology is built has provided this generation with lots of options. Having options makes them feel as if they were making the right decision for themselves that best suit their needs. No matter if it is the new mobile phone or a new watch is out, the first thing teenagers will ask is how many options does it have and what can it do? Being familiar with this type of behaviour also is extended to their everyday life interaction with other humans, namely their parents, teachers, and other people who should also give them options.

Like things customized

Everything should be customized to who they are and to the way they use it and this goes for fashionable clothing as well as advanced technology gadgets. It made them feel unique because that object was made especially for them. In the world where there is mass production, people tend to desire to be unique.

Learn and create one's own digital language

You can see shorter form of words that look like slang or funny symbols practically came up in every online conversation. The Thai teenagers use this type of language in chat rooms and when they send messages to each other through MSN. The most common reason why they want to create this type of language is due to it is easier to communicate with others because you type less. It is understandable only within this new generation giving them the sense of secrecy, hip, being in the "in" group, and being different. The new types of language are in two distinguished forms: symbols and slang. Most of the words came from slang teenagers use with friends in real life, sound imitations, or abbreviations. Below are examples of new types of language.

Symbols	*Meanings*
T_T	Sad or crying
- _ -"	So tired that you sweat or didn't understand
^_^	Happy
555+	Laugh
lol	Teasing, sticking the tongue out or laughing at
>o<	Shouting out
^o^	Laughing
: P	Teasing, sticking the tongue out or laughing at
- _ - #	Stressful
…	Don'thave anything to say
English Words	*Meanings*
D or ddd	Sawaddee, Hi in Thai
Bub Bye or by	Bye bye
G"nite or Gd nite	Good night
2 U	To you
ya	You
Chillin" or chill chill	Just hanging out, a "so-so feeling"
thx	Thank
plz	Please

Thai Words	*Meanings*
ซุย	"Sui" means stating that one is boasting or bragging about things
ซะงั้น	"Sa-ngan" means "why did it happens this way"
เล่นเอ็ม	"Lend-m" means playing MSN
กำ	"Khum" means bad karma, paying one's bad karma
ก๊ากกกกกกกกก	"Kakkkkkkk" means very funny
เม้น	"Ment" is an abbreviation of "comment"
หุหุ , หึหึ , กิกิ, or อิอิ	"Hu-hu," "Heu-Heu," "Kik-kik," or "Ai-ai" means laughing

Form new types of relationships

Many new types of friendships are formed through the help of advanced technology like the internet. Unlike writing a letter to a pen pal, correspondence through chatting, MSN, ICQ, QC, and e-mail can establish a much faster friendship and a larger group of friends with the same interests all at once. Establishing a romantic relationship through the internet is also something these teenagers experience. Many types of relationships have evolved side-by-side with technology advancement. For example, the new form of fan club is another type of modern relationship. Fan club activities in the computerized generation take shape in a new way, besides sending a letter to their favourite celebrity. Teenagers who like the particular celebrity may make a website so that others who like the same celebrity could join in. Activities in this kind of web site includes updating pictures of their favourite celebrities, writing the biography, and posting on the web the events when that particular celebrity will show up. Similar activities can be found on television too. Reality television also provides a new type of participation for the fanatic teenagers to send messages for others to see on screen and vote for their favourite stars.

Learn new skills

The usage of computer provides teenagers with direct and indirect learning skills. The skills these teenagers stated that they learned are such as English practicing skills, social and conversational skills, typing, screening, searching for information, data entering, analyzing information, using Microsoft programs namely Word, Excel, Power Point, editing pictures through using arts and Photoshop programs skills and problem solving, the skills of making a website, and even playing computer games provide them with finger and eye coordination skills!

Addiction to new things

Like drugs, teenagers can be addicted to many things such as addicted to playing games, addicted to the internet, addicted to chat rooms, and addicted to friends.

Setting new etiquette "the netiquette"

Being in the internet world also requires a person to behave within the netiquette rules. These Thai teenagers who often use online service such as chat and posting comments on web board know these shared rules. For example, a person is prohibited to say or post rude words, hack, stole data, to capitalized every letters because it implies that you are shouting rudely to the others, post inappropriate pictures on the net, forwarding junk mails, sending unrelated e-mail to everyone on the e-mail list, and respect other people opinions and the website rules.

These are just the main characteristics that computers have on the computerized generation. There may also be other characteristics that are unique to just a small group of teenagers. Moreover, not every teenager will be influenced in all the ways that are listed above. Some may be influenced more than others. It also depends on the individual's personality. For instance, someone who is an extrovert will not find it necessary to be someone else online and use multiple identities. Furthermore, someone who is an introvert will have limits in expressing his/her emotions even if the person on the other side of the screen does not know who they are. However, the majority of the characteristics are interrelated. One will lead to another and so on.

Adults' False Myths Concerning Teenagers and Computers

During the interviews, teenagers also mentioned that adults in Thai society often have some negative connotations and false myths about computers.

"A new creation, like computers, always has a double standard on humans."

The new generation is aware of the fact that everything has two sides. It is up to the user's sense of responsibility which will make things worst or better, but not the machine.

"Computers are luxurious urban accessories that will partly lure kids to get addicted to bad things such as pornography and games."

It is not always true that how these teenagers use computer would turn out to have a bad impact on their lives. Teenagers think that computers are not a luxury machine any more because they use them in a very down to earth way.

"Computers are best to be used as working machines and office supplies or for educational purposes rather than other usage."

Teenagers think that computers are best to be used in the most optimized way—that is it should be used in various ways than just to use it for working and direct learning purposes. This is where the real value of computers lies. A lot of computer-related activities that teenagers engage in teach them valuable lessons in life, help them grow, and most of all those indirect learning experiences are hard to be found elsewhere when both of their parents are working outside the home. In order to survive in this new world (computer and high technology world) they need those skills.

"Adults think that we (teenagers) only use computers for entertainment and fun"

This is not always true. Teenagers use their computers for multi-purposes. Although most of the time is for entertainment-related activities, they also use computers to educate themselves by searching for new information, finding their self-identity, and learning indirect skills (using English vocabulary, typing skill, etc.), which all are beneficial to them.

"Computers block teenagers from the outside world of activities."

This might be true if the term "outside activities" equals to physical activities such as playing sports and being in the nature surrounded by environment. But for these Thai teenagers, computers open up another window of opportunity for their generation that their parents never had and wish to have. To them, meeting new friends, exchanging opinions with people across the world, and forming their own community which composed of people who share the same interest are their meaning of "outside activities." As time passes, adults might have to compromise to get used to the new form of "outside activities."

"Chatting on the net is nonsense and it is just a waste of time."

Adults would be surprised that often teenagers use chat for beneficial reasons. They make new friends, ask how to do homework, where to search for good source of information for doing reports, conducting their small research project asking how the people in the chat room think, and helping them make the right reasons by consulting the others who have experience.

"Computers can bring harm and danger to children and teenagers."

Although this can be true, teenagers think that their generation is aware of this fact. Their generation has created self-mechanism ways in protecting themselves from harmful things. For example, when they chat on the internet, they will not tell their personal information to other strangers. If they think there is something fishy, they will consult their parents. They know how to balance their time between playing computers and working/studying. They do not believe in everything they read on the internet. They will use their own judgment on the things they read. Last, they know that playing computers too much will make them short-sighted; therefore, they should keep a one foot distant from the screen and rest their eyes when they have played computer for more than two consecutive hours.

"Computer games make your child violent."

Teenagers think this is not true. They think what lead kids of their age to be violent is the fact that some kids do not have proper parental guidance, do not have good role models at home, and kids that lack parents' love and affection are those who act violently towards others. These reasons are likely to shape teenagers to act aggressively and violently.

"Computers are too high tech for adults to learn and keep up with. It is just a machine that has no feelings."

Teenagers' advice to this misconception is for adults to open up their minds. Learning new programs is hard for any age. It takes time, patience, and most of all a sense of nurturing toward your computer to booth up your attitude and help you to learn faster than before.

In sum, teenagers tend to view computers in a more positive way than adults. One of the reasons these teenagers think that cause adults to have these negative notions is the news. From time to time, there will be negative news on television and in newspapers examples of computer and internet danger, which happens in Thai society or the Western countries. These examples are such as how some of the Thai children and teenagers skipped school to play online games, playing computer too much causes the children's grades to drop, children learning about sex and are likely to desire to engage in sexual activities earlier than they should because of viewing pornographic pictures on the internet, how crimes like kidnapping and luring girls to be raped can happen through internet chat, and the most common one was children can imitate violent acts they see on their computer games in real life.

How We Foresee Our Future Living with Computers and Advanced Technologies

A lot could be learned about one's future through listening to the imagination of others. A part of this study emphasizes on valuing the imaginations of this young generation who will soon grow up to be future adults, the important source of the country's brain and strength. Often, understanding the imagination of others is a difficult task. With the help of pictorial ethnography techniques, it can help bridge the understanding to the imagination description given.

Thai teenagers visualized their future living with computers and advanced technology in similar ways. We would live in a world where computers control everything. We would live a comfortable life where things are done much faster and more conveniently. Every household, even the poor, will own at least one computer. Computers have now become our life.

They see themselves living in a world full of another citizen, robots. Robots are like the fifth means of livelihood after food, clothes, shelter, and medicines that humans could not lack. Thai people would start preserving antique Thai style houses. People will go back to learn more about their traditional religious beliefs to uplift their spirits. Inside our house, we will use computers to control every type of housework. Robots will be our maids and do the tasks housewives used to do. The time we save from not having to do these chores will be used to do other things that we want in life. Every function we want in daily life can be done with the use of our mobile phones or wrist watches.

Education will improve a lot due to technology that helps us in visualizing what we learn much more efficiently than reading books. We will develop an advanced technology to improve the visual and audio system. It is the responsibility of the local government to added new type of welfare for children such as lessons through the internet. In the future children would not have to go to school. They will use the internet instead. They can search for information directly and if they have problems they will e-mail their teachers. Every message such as chat messages, school reports, e-mailing friends on the other side of the world could be possible through the internet technology. Parents should learn to support their children use of technology in a proper way. A part of having good parental skills will depend on the level of technological knowledge each parent has and how they teach their children to use them in the right way. Parents should also encourage their children to exercise to keep them healthy.

On the street we might see computers replacing merchants and street vendors. We will not have to drive our car, it will just move according to how we program where we want to be. We will also travel in the air. There would be traffic control in the air as well as on the ground and underground. Time will be very valuable; and therefore, it is the most important thing for human. We cannot waste anytime because and everything move in fast pace. Everyone is in a rush, time spent making each decision is crucial because we will live in a more competitive world. We will use transportation that is fast like the sky train. Cars will also be limited due to it can cause lots of pollution. Thus, most people will have to use public transportation like the sky train and subways.

Food will come in a more processed form, maybe in a can. Canned food will be composed of all the essential nutrients humans need including added vitamins and minerals. When we are sick we can cure ourselves. There will be no need to go to the hospital and spend an hour waiting for doctors. There will be self-used medicine in every household cabinet.

Our physical body might change too. Our head might be larger than other parts of the body. This is because the brain is being used the most. We lack the exercise of using our arms, legs, etc. We might have computer chips installed in our body. The chip will act like a remote control that will control the brain waves that can send messages to the internet network. The chip can also link to computers at home, mobile phones, and other electronic devices that we use. Humans will become lazy as everything will be done using technology. We will have more illnesses and diseases because our health will get worse as we lack exercise from doing things for ourselves. We will be sitting a lot, become short sighted, and fatter because we rely on robots to do everything for us.

The downside to the story is the fact that industrialization and advanced technology have brought along many bad impacts on the environment. They destroy the water, the air, the soil, and the ecology system that we live in, causing rapid growth of various types of pollution. We have used up all of the natural resources. However, humans have found a way to create a new type of technology that will help build a new ecology system that we long for. It is a shame that our future might be limited to the tight space we have created. Nevertheless, people can travel through different periods of time using the time machine and can live in the time that they want. It becomes a cycle where old technology will be used to conduct endless research in the desire to make new advanced technology that is better and can replace the old ones.

Technology will help us fight crime, catch burglars, and discipline people who break the law. We then might have only one government that is the world government. People would have to follow the rules from this organization. We as humans, no matter what race, ethnicity, gender, age, or class, will have to live under the same rule that is "the world law." Overall, it is still a good world to look forward to.

What these teenagers described not only reflects the picture of their future, but also reflects tomorrow's living condition based on what they experience in today.

Ethnographic pictorial stories of computerized generation's future lives

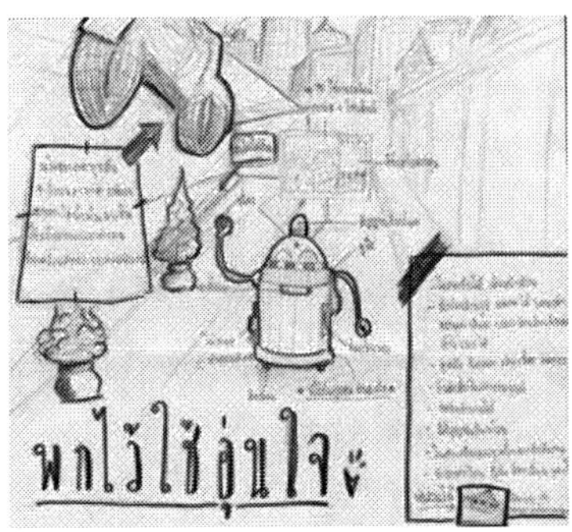

Conclusion

The conclusion of this research does not aim at making generalisation on the whole population of the Thai computerized generation but rather aim to give insights into how this particular group of Thai teenagers in the study perceived and interacted with computers. The way this sample population define themselves gives us a clue on how they distinguish themselves from other generations in Thai society. The description of their daily life activities highlights the way they utilize computers and advanced technology. Symbolic meanings given to computers by this computerized generation provide us with a deeper understanding of "why" computers have become an essential part of their lives; whereas, unique characteristics of this very same generation provide us with a better understanding of "how" computers become their ways of living. Living under the globalization trends might have made some of the concepts mentioned as characteristics of the Thai computerized generation seemingly similar to what has been found in Western countries. When one looks closely and digs deeper from the surface, one will find that all of these influences are new to the Thai society. Many of the computer influenced characteristics cause hybridization of the ways these Thai teenagers adopt technology into their lives. Furthermore, my decision in including teenagers' voices on the topics of "Adults' False Myths Concerning Teenagers and Computers" and "How We Foresee Our Future Living with Computers and Advanced Technologies" serve as an encouragement and an opportunity for these teenagers to have a say on issues which are relevant to their presence in the Thai society. In this study, I have mapped out different aspects in studying the lives of the Thai computerized generation. Future studies could use this research as a foundation in exploring variations of technological and computers-related impacts on those of different generations, genders, ethnicities, ages, classes, and religion groups in Thai society. Such studies can be put to use for serving the different needs of each group in the society.

References

Anderson, Ben and Karina Tracey. 2001. Digital living: The impact (or otherwise) of internet on everyday life. *American Behavioural Scientist 45*(3): 456-475.

Chaddatar Paratasillapin. 2001. *Internet use of the pre-computer generation users.* M.A.Thesis. Faculty of Communication of Art, Chulalongkorn University.

Dooling, Judith O"Donnell. 2000. What students want to learn about computers. *Educational Leadership* (October): 20-24.

Internet users' behaviours. 2003. *Econnews* (April): 47-49.

Khanchit Malaiwong, et. al. 1998. *The impact of technology on Thai bureaucracy: A research report.* Bangkok: The National Electronics and Computer Technology Center.

Komonmas Paradonprateave. 1991. The evolution of computer and computer used for educational reasons in Thailand. *Sukothaithaamatiraj Journal 4*(2): 1-17.

Monchai Sripetchnai. 2004. *Factors impact on internet behaviour among teenagers in Bangkok.* M.A. Thesis Communication of Art, Thammasart University.

Ongarij Ritthongpitak. 1996. *The use of world wide web on internet among university students.* M.A.Thesis. Department of Public Relations, Chulalongkorn University.

Ornuma Srisuttipan. 2002. *Opinion, media exposure and lifestyle of internet users in Bangkok.* M.A. Thesis. Faculty of Communication Arts, Thammasart University.

Pentip Jirapinnusorn. 1997. *Searching for information through mass media and internet among college students and university employee.* M.A. Thesis. Faculty of Communication of Arts, Chulalongkorn University.

Phetchompu Theppipit. 1990. *The relationships between violent behaviours and behaviours of reading cartoon books: Viewing television shows and playing computer games of the six grade students in Bangkok.* M.A. Thesis. Department of Telecommunication, Faculty of Education, Chulalongkorn University.

Pituma Puntawee. 1999. Guess...Who use the internet. *IT Pritat* 7(6): 7-11.

Piyanat, Pondper 1999. Peaking at the internet user behaviours. *E-commerce* (December): 49-50.

Sarawut Puntuwong,. 2003. *Receptive behaviour: Awareness and utilization of Thai-web sites from teenager in Bangkok area.* M.A. Thesis. Faculty of Communication Arts, Thammasart University

Sirintorn Chaisakda. 2002. Survey result of internet user in Thailand in the year

2001. *IT Praritat 10*(3-4): 5-8.

Srichan Changwattanachai. 1991. The roles of computer in Thailand. *Saamitsan*(July-August): 89-94.

Srisab, Punsak. 1999. Internet and the elderly people. *Computer Today* (September): 92-97.

Tapscott, Don. 1998. *Growing up digital: The rise of the net generation.* New York: McGraw-Hill.

Thanat Kaseamchaiyanand. 2001. *The factors that effect web site accessing behaviour of internet users in Bangkok.* M.A.Thesis, Faculty of Communication of Arts, Thammasart University.

Tiotuyco, Gonnzalo P. and Pakorn Vichyanond. 1985. Computerization: It all began
with a pair. *Thailand Business* (September): 36-44.

The National Electronics and Computer Technology Center. 1999. *How the Thais use their internet?: A survey report.* Bangkok: The Center.

The National Electronics and Computer Technology Center. 2004. The internet index of Thailand. [http://www.ntl.nectec.or.th/internet/user-growth.html]. 12 December 2004.

The National Electronics and Computer Technology Center. 2002. *WWW: Thaiparents.net.net.* Bangkok: The Center.

The survey of information technology: Report results on the usage of advanced technology. 2003. Bangkok: the National Electronics and Computer Technology Center.

Williams, M. Samuel and Leighton C. Peterson. 2002. The anthropology of online communities. *Annual Review Anthropology 31*(1): 449-67.

CHAPTER FIVE

INFORMATION ETHICS: LOCAL APPROACHES, GLOBAL POTENTIALS? OR: DIVERGENCE, CONVERGENCE, AND ETHICAL PLURALISM AS MAINTAINING DISTINCTIVE CULTURAL IDENTITIES *AND* (QUASI?)-UNIVERSAL ETHICS

CHARLES ESS
CENTER FOR INTERDISPLINARY STUDIES, DRURY UNIVERSITY

Introduction

The ongoing, exponential growth of the Internet and the World Wide Web continually expands instantaneous linkages between an ever-increasing range of very diverse cultures. At the same time, Information Ethics and Computer Ethics seek to examine and resolve the ethical issues evoked in the design and deployment of Information and Communication Technologies (ICTs), utilizing fundamental value commitments and ethical decision-making approaches—which vary from culture to culture. The realities of radically diverse cultures now wired together on a global scale thus present us with a central set of issues: how may we develop Information Ethics and Computer Ethics that

(a) address both local and global issues evoked by ICTs / CMC, etc. ,
(b) in ways that both sustain local traditions / values / preferences, etc. and
(c) provide (quasi-) universal responses to central ethical problems?

Of course, these problems are not especially new: rather, they are as old as the efforts in the Western philosophical traditions to discern universal, or at least quasi-universal ethical norms that could claim legitimacy beyond the boundaries of a single cultural/ethical tradition. What is new here is that the Internet and the

Web bring to the foreground the *reality* of cultural diversity in dramatic new ways, as these technologies provide linkages between cultures across a scope and with a speed unimaginable in earlier times—and thereby demonstrate the increasingly urgent need to determine whether or not a global Information Ethics and Computer Ethics are indeed discernable. That is, whereas in earlier eras it was easier to remain provincial, if not ethnocentric—the Internet and the Web confront us with the reality of cultural and ethical diversity, and thus transform what might have been an interesting academic question into an urgent real world task.

In the following, I argue that one possible resolution to this task is the development of *ethical pluralisms* that conjoin shared ethical norms with diverse implementations and understandings of these norms, as shaped by diverse cultural traditions, histories, values, and worldviews. I first review a series of examples in the domains of Information Ethics and Computer Ethics, emerging in the Western world, that foreground such pluralism in transnational information ethics. But while these pluralisms can be seen in emerging ethical guidelines for research that straddle considerable cultural differences—these first examples remain clearly within "the West."[24] I then turn to recent developments in Information and Computer Ethics in Asia. Here again, we find that the even greater cultural and ethical differences marking Western and Eastern approaches—most centrally, to issues of privacy and data privacy protection—nonetheless likewise allow for an ethical *pluralism* that holds together important ethical norms alongside the irreducible ethical traditions and approaches that define diverse cultural identities around the globe.

The ethical geographies of recent efforts towards interdisciplinary / international approaches to Information Ethics

Association of Internet Researchers ethical guidelines

As I have described elsewhere (Ess 2003), the Association of Internet Researchers (AoIR) has developed a set of ethical guidelines for online research (AoIR 2002). Such research realized early on precisely the ability of the Internet and the Web to bring together not only researchers from a variety of disciplines, but also whose ethics were shaped by a considerable diversity of national ethical

[24] Of course, post-colonial scholarship has effectively demonstrated that the terms "East" and "West" are the artifacts of colonialism and imperialism, and as such, their meaning and use is problematic for a range of reasons. Nonetheless, I use the terms—in full recognition of their history and limitations – simply as a useful shorthand.

traditions. Indeed, to our knowledge, the guidelines approved by the AoIR membership in 2002 represent in the English-speaking world the first interdisciplinary, international set of ethical research guidelines.

A central strategy in developing the AoIR guidelines was to determine whether or not we could find *pluralistic* middle grounds between the familiar poles of ethical relativism and ethical dogmatism. We may define ethical dogmatism as the view that there is only one set of values, universally valid for all peoples in all times in all contexts—and so differences in ethical values and practices can only mean that one set is right and the others are wrong. By contrast, ethical relativism claims that there are no universal values whatsoever, and so any set of values and practices are as good as any other—all values are legitimate only relative to a given individual and/or culture. As a middle ground, ethical pluralism points to an agreed-upon value or belief (in the first example we are about to see here, on the *expectations* regarding privacy that become normative with regard to requirements for informed consent). At the same time, however, pluralism emphasizes that different cultures may interpret, apply, and/or understand these norms in different ways that reflect—and thereby preserve—each culture's distinctive history, traditions, values, etc.

Early on in the work of the AoIR ethics working group, we encountered a clear difference between the United States and Norway regarding requirements for informed consent for recording (e.g., audio and/or video) in public spaces. In the U.S. context, as Joe Walther documented, no informed consent for such recordings are required—first of all, because there are in this context no *expectations* of privacy, in contrast, for example with conversations in a psychologist's office, etc. (Walther 2002). But informed consent for such recordings in Norway is in fact required—precisely because people do not expect such recordings without consent (Elgesem 2002). Clearly, while *expectations* regarding privacy *differ* between the U.S. and Norway—there appears *agreement* here that these expectations are nonetheless a primary starting point for making ethical judgments as to whether or not informed consent is required. In this way, we see a *pluralistic* structure of *agreement* on the primary normative importance of expectations alongside *diverse* expectations reflecting distinctive national / cultural differences. While a (quasi)-universal standard (*expectations*) emerges—it is one that works to preserve important cultural / national differences.

Similar pluralistic structures can be seen in the RESPECT project, which sought to develop ethical guidelines for socio-economic research across the European Union (see <www.respectproject.org>). Even earlier, Paul Reidenberg has argued that despite considerable differences between E.U. and U.S. approaches to data privacy protection, these approaches reflect diverse ways of seeking to implement shared "First Principles" of Data Protection. Very briefly, the European Union pursues policies that recognize the role of the state to protect individual rights—including privacy and data privacy rights as rooted in fundamental human rights—through legislation (Reidenberg 2000, 1331f.). By contrast, the United States relies primarily on self-regulation rather than governmental regulation, seeking to protect privacy "...through practices developed by industry norms, codes of conduct, and contracts rather than statutory legal rights. Data privacy becomes a market issue rather than a basic political question, and the rhetoric casts the debate in terms of "consumers" and users rather than "citizens"" (1332). (In other terms, this represents a characteristically *consequentialist* position, one that emphasizes economic benefit at large over possible risks to individual privacy: cf. Burkhardt, Thompson, and Peterson 2000, 329.) Again, this appears to establish a *pluralistic* structure in which *agreement* on first principles stands alongside distinctive cultural / national *differences* regarding how to implement and apply these first principles.[25]

NESH (National Committee for Research Ethics in the Social Sciences and the Humanities, Norway)

As a last example of such pluralism, we can consider the requirements for informed consent and protection of research subjects' confidentiality and anonymity as these operate in the United States and Norway. The Norwegian guidelines for research ethics (NESH 2001) are striking first of all for their emphasis on the *intrinsic* value of knowledge as the primary reason for its pursuit—in contrast, that is, the more utilitarian emphasis on the potential *benefits* of knowledge characteristic of similar documents in the U.S. and the U.K. (e.g. Code of Federal Regulations). Moreover, the NESH guidelines emphasize that researchers are obliged not simply to respect and protect the individual, but also close relations thereof (see para. 40). This more *communitarian* understanding sharply contrasts with the emphasis in U.S. and

[25] Similar convergences can be seen within the fields of Information Ethics and Computer Ethics *per se*. As I have documented elsewhere, Terrell Ward Bynum has argued that in these fields there is a convergence in the work of Norbert Wiener and Luciano Floridi – one oriented towards the central values of: contributing to human flourishing; advancing and defending human values (life, health, freedom, knowledge, happiness); and fulfilling "the great principles of justice" drawn from Western philosophical and religious traditions. Bynum further sees agreement on these central values in the work of such computer ethics pioneers as Deborah Johnson, Philip Brey, James Moor, Helen Nissenbaum, and my own emphasis on using computer-mediated communication (CMC) technologies in ways that preserve "thick" or local cultures (Ess 2005c).

U.K. ethical codes on protecting solely the individual.

But these sharp contrasts once again appear to resolve into a pluralistic structure that holds shared agreements alongside irreducible cultural and ethical differences. That is, in both the U.S. and Norway, there is agreement on the primary importance of basic human subjects' protections in research, including rights to privacy, confidentiality, anonymity, and informed consent. At the same time, however, there are clear differences in how these protections are understood—reflecting larger national differences between U.S. *individualism* and Norwegian *communitarian* emphases on the well-being of persons and their web of close relationships.

Of course, I do not mean to argue that all differences between diverse national and ethical traditions can be resolved into a tidy structure of pluralism. On the contrary, even when such pluralisms can be discerned in some areas—differences remain that can result in profound ethical and legal conflicts nonetheless.[26] Still, these examples of ethical pluralism strongly suggest that cultural diversity in approaches to Information and Computer Ethics does not force upon us the simple dualism of either ethical relativism or ethical absolutism. Rather, ethical pluralism emerges here as a middle ground that points to shared norms alongside distinctive approaches to understanding and

---

[26] We will see one example of such conflicts below, in the contrast between the research ethics at work in the Yasukawa and Ando study from 2002 and the AoIR ethical guidelines. An even more striking contrast is at work in an example I have documented elsewhere. Briefly, a sharp contrast emerges between the ethical assumptions underlying an online game, designed in an Asian country for the sake of determining customer preferences for next-generation network services, and research ethics and privacy law in Western countries, specifically Denmark. For the Asian designer, the game is morally unproblematic because it collects no personal information – defined to include IP addresses, personal names and addresses, etc. From the standpoint of Danish data privacy law, however, the demographic information collected in the game (age, gender, income) is included *within* the definition of personal information – and thus requires the basic protections of informed consent and guarantees of privacy and confidentiality (*Persondataloven* 2000, 2001). Moreover, the game design is morally problematic from a Western standpoint because it fails to inform participants of the purpose of the game; nor does it ask for informed consent for using the demographic data provided by users; and the design has no procedure for protecting this data.

Beyond considerable differences between Western and Asian countries and traditions regarding privacy (explored more fully below), the contrasts here may further reflect the contrast between Western conceptions of rights as focusing on the *individual* as rights-holder, and Asian traditions (especially as influenced by Confucian thought), as these emphasize group consensus and harmony as primary (Ess forthcoming). In any event, this example makes clear that whatever pluralisms may emerge between elements of Asian and Western Information Ethics and Computer Ethics – again, such pluralisms will not eliminate all conflict and disagreement.

implementing those norms in ways that preserve and foster cultural diversity and identity.

At the same time, however, it can be objected that these examples of pluralism all occur within a Western domain. Especially as use of the Internet and the Web in Asia now exceeds that of North America—it is clear that our computer and information ethics, however, will have to develop a genuinely global extension, one that will extend between East and West. Can structures of pluralism work to encompass both important agreements and the even greater cultural differences between Western and Asian cultures and traditions?

East-West approaches to Information Ethics

In fact, we can see a number of such pluralisms across East-West cultural differences nonetheless emerge in Information and Computer Ethics in several ways, beginning with a more theoretical approach to the possible philosophical foundations of such ethics.

Confucius and Aristotle as foundations for East-West Information Ethics?

A first, if general, example of pluralism *between* Eastern and Western ethical traditions can be seen in the multiple resonances explored especially in comparative philosophy over the last few decades between Aristotelian and Confucian thought. As I have discussed more fully elsewhere, these resonances begin with similar conceptions of *human nature*—and extend from there to similar understandings of the primary aims of *ethics* and *politics* (Ess 2004a, 2004b, 2005b).

So, to begin with, both Aristotle and Confucius emphasize our interconnectedness with one another in the human community. Aristotle states in his *Politics*: "man is by nature a political animal," so that "… he who is unable to live in society, or who has no need because he is sufficient for himself, must be either a beast or a god: he is no part of a state" (1943 54f./1253a). Aristotle develops from here a *communitarian* tradition in ethics and politics—one further represented in both Kant and contemporary Habermasian discourse ethics (cf. Ess 2005a). This interconnectedness finds its counterpart in what Ames and Rosemont characterize as a processional or relational ontology in Confucius, i.e., "not a concern to describe how things are in themselves, but how they stand in relation to something else at particular times" (1998, 23). This means in turn that "Persons are not perceived as superordinated individuals—as agents who stand independent of their actions—but are rather ongoing "events" defined functionally by constitutive roles and relationships as they are performed within the context of their specific families and communities ..." (1998, 29). Confucius thus directly anticipates Aristotle: "...one cannot become a person by oneself—we are, from our inchoate beginnings, irreducibly social" (1998, 48).

By the same token, central components of Confucian ethics and politics likewise anticipate Aristotle as they emphasize *harmony* in a community, beginning with the family:

Ancestor reverence as the defining religious sensibility, family as the primary human unit, authoritative humanity (perhaps more literally, "co-humanity," *ren = human or mankind*) and filiality (*xiao*) as primary human values, ritualized roles, relationships, and practices (*li = power or energy*) as a communal discourse, are all strategies for achieving and sustaining communal harmony (*he = harmony*). (Ames and Rosemont 1998, 30)[27]

In particular, Tu Wei-Ming argues that Confucian ethics resonates with Aristotle's virtue ethics as well as with contemporary feminist and environmental ethics. Similar to Western ethics of care for both the human and ecological communities, a Confucian *ethos* stresses fidelity and fiduciary responsibility to a community that begins in family and finally encompasses the world (Tu 1999, 35).

These basic and very strong similarities, however, do not obscure irreducible differences between Aristotelian and Confucian approaches. To begin with, Aristotle's ethics and politics, as part of the *practical* aspects of human reason and experience, are to be conjoined with a *theoretical* knowledge that, in the long run, approximates a God-like comprehension of what we might think of as the scientific laws of nature. While Chinese philosophy certainly included careful attention to nature, issuing in important theoretical and technological insights and discoveries—at least for Confucius, such concerns are far less significant than they are for Aristotle.

But alongside these irreducible differences, marking off these traditions from one another, we can further discern nonetheless a now familiar structure of pluralism—one that sees important coherencies and convergences in Arisotelian and Confucian thought with regard to basic conceptions of human nature, ethics, and politics. Especially as these Aristotelian-Confucian resonances may serve as one element in the development of a global information ethics, as I have argued (Ess 2004b), such an ethics would rest precisely on a *pluralistic* approach that holds important agreements alongside the irreducible differences defining diverse cultural and historical traditions.

[27] My thanks here to Profs. ZHANG Ning and NAKADA Makoto for their help with determining the Chinese characters at work here.

On the ground... contemporary discussions

Beyond these more theoretical reflections, we can see more particular sorts of pluralism arise in examples drawn from Japan and China. From here, I will then explore a pluralism between Western (U.S. and German) conceptions of privacy and data privacy protection laws and emerging conceptions of privacy and data privacy protection laws in Asia.

Internet research ethics in Japan?[28]

Briefly, Information Ethics has developed recently in Japan, especially under the aegis of the "Foundations of Information Ethics" (FINE) project (see <http://www.fine.bun.kyoto-u.ac.jp/index_en.html>, and is represented in English-language literature, for example, by the work of Masahiko Mizutani and his colleagues (e.g., Mizutani et al 2004). While there exists no Internet Research Ethics (IRE) *per se*, several of the basic components of IRE are in place, including copyright law and requirements to respect privacy and related human rights in research.

Nonetheless, there are clear differences in approaches to IRE in Japan vis-à-vis Western approaches (as articulated, for example, in the AoIR guidelines)—in part, perhaps, because of cultural differences. For example, in one recent study of an online therapy group, the researchers used direct quotes but without informed consent—believing, apparently, that the identity of the participants was sufficiently protected by using pseudonyms (Yasukawa and Ando 2002, cited in Tamura 2004). Such an approach would *not* be endorsed by the AoIR guidelines: on the contrary, these guidelines explicitly warn against using direct quotes taken from message boards and other publicly accessible resources because the authors are trivially easy to track down through a string search with Google or similar search engine.

For his part, in his more recent study of messages exchanged in a forum, Tamura took what he characterizes as a "more cautious way." This more cautious way including requesting consent for the use of paraphrases, not direct quotes, of forum messages, coupled with pseudonymizing the authors (Tamura 2004). This approach is more directly consistent with the approach recommended in the AoIR guidelines—and this despite clear differences between the (largely Western) conceptions of privacy underlying the AoIR guidelines and those at work in Japanese culture.

---

[28] I owe virtually all of the material here regarding information ethics in Japan to the work of Mr. TAMURA Takanori, primarily, his unpublished report (2004), along with subsequent discussions with Mr. Tamura and his colleagues, including Professor NAKADA Makoto, Mr. TAKENOUCHI Tadeshi, and Ms. Leslie-Tkach Kawasaki.

Very briefly, Nakada et al (2004) have argued that Japanese culture includes several components relevant to conceptions of privacy, beginning with *seken*. *Seken* is understood as a primary "world"—a complex of relationships and way of life. In this light, a naïve person is one who does not know *seken*. By contrast, the cultural domain of *shakai* entails a cluster of values and notions largely imported from the West, including notions of democracy, individualism, society, law, free speech, and human rights. On this analysis, a primary notion of privacy in Japan is first of all associated with *shakai*—along with *positive* valuation of the individual and the self. By contrast, *seken* is deeply influenced by Buddhism, including the importance of no-self" (*musi*). In *seken*, *denial* of self is a virtue—expressed, for example, in the practice of *revealing* secrets to others as a way of confessing one's impurities but also thereby developing friendships.

Clearly, there is a strong *contrast* between here between Western emphases on the *individual* and thereby on privacy and privacy protection as a good (*shakai*)[29] and what we might think of as a more *communitarian* valuation of personal disclosure as a good (*seken*)[30]. At the same time, however, there is certainly interest in Japan—expressed in recent law—to protect privacy, especially data privacy. This suggests again a *pluralist* structure, one that balances *agreements* on the importance of privacy and ways to protect it online, alongside clear cultural differences in conceptions of the individual and privacy.

---

[29] I will develop Western understandings of privacy more fully below.

[30] These associations between different notions of privacy and the cultural spheres of *Seken* and *Shakai* are confirmed by more recent research findings by Prof. NAKADA Makoto, TAMURA Takanori, and Leslie Tkach-Kawasaki:

1) Japanese people (the respondents) are aware of the importance of privacy in many ways.

2) But this awareness of privacy tends to be related with *Seken*-related meanings.

For example, "Japanese way to get good friends by betraying or sharing secrets or information on one's privacy" or "criticism of spam-mail as sources of invasion of privacy" is related to (some of) *Seken*-related meanings.

3) On the other hand, "individualism" is negatively related to awareness of privacy as well as to interests in political problems and in environmental problems, so far as the tentative analysis on our research data shows.

Professor Nakada observes:

Generally speaking, these findings are consistent with the contents of our discussions about privacy. We, Japanese and "Western" people, share common attitudes towards evaluations of privacy but these similar attitudes seem to be derived from different cultural backgrounds; individualism and *Seken* (or "collectivism" related to *Seken*) (Nakada 2005).

Information Ethics in China

Professor Lü Yaohuai has written the first book in Chinese Information Ethics (2002).[31] Here he explores in particular a number of concerns raised as globalization and the Internet and the Web dramatically expand access to the rest of the world for China and her people. On the one hand, Prof. Lü argues for the need for a universal ethics in global ICTs—while further recognizing the often deep differences between Western countries and the information ethics they have developed, on the one hand, and, on the other hand, an indigenous Chinese information ethics that reflects especially Confucian and Buddhist traditions. He suggests here that the neo-Confucian notion of *shen-du* (i.e., *du* = self, *shen* = regulation or self-control) might be a useful ethical resource, in light of the fact that otherwise, neither law nor ethics can prevent people from engaging in immoral activities online.[32]

Interestingly enough, this notion of self-regulation offers a striking parallel to similar approaches to research ethics in a range of Western countries, such as Germany, Denmark, and Sweden. In these countries, there are no "Institutional Review Boards" (U.S.), "Research Ethics Boards" (National Research Council, Canada), external Learning and Teaching Support Networks' subject centres and internal Academic Standards and Policy committees in the United Kingdom (U.K.), the National Health and Medical Research Council (Australia) and the Australian Research Council, etc., as shaped by [U.S.] responses to human rights abuses in medical experiments and social science research. Rather, *researchers are trusted to do the right thing*—and if not, it is (sometimes) possible to complain to national boards or committees (e.g., the Danish Research Council for the Humanities) who may issue comment.

It remains to be seen, of course, on whether (and if so, how far) *shen-du* will gain currency as a component of information ethics in China. But its status as a proposed approach to information ethics in China suggests at least a possible pluralism between Western notions of self-regulation and their counterpart in Chinese information ethics.

---

[31] My access to this volume is thanks to the translation efforts of my students at the University of Trier during Winter Semester, 2004, Na Yang, Wen Li, and Chenchen Song. I am very pleased to be able to express my gratitude to them for their hard work and assistance in making portions of Professor Lü's volume accessible to me.

[32] Again, my thanks to Profs. ZHANG Ning and NAKADA Makoto for their help with determining the Chinese characters at work here.

Privacy in U.S. / Germany / Hong Kong / China—a pluralistic phenomenon?

As Johnson (2001) points out, justification for privacy in especially the U.S. begins with some sense of privacy as an intrinsic good—i.e., a good in itself that needs no further justification. Moreover, privacy serves as an instrumental good, i.e., as a means to develop a sense of self and personal autonomy, intimate relationships, and other capacities and abilities. In addition, privacy funds the basic elements required for participating in a democratic society—i.e., personal autonomy/freedom and then the capacity for dialogue, debate, etc.

If we turn to Germany, we find similar rationale for privacy, beginning with a notion of privacy as a basic right of an autonomous person *qua* citizen in a democratic society. Privacy is further instrumental in the protection of private autonomy, the freedom to express one's opinion, the "right to personality" (*Persönlichkeitsrecht*), and the freedom to express one's will. In addition, however, privacy protection—specifically, data privacy protection—is seen as a means necessary for the development of e-commerce (Bizer 2003).

It is this latter rationale that we then find at work in Hong Kong and China. According to Tang (2002), the rights to data privacy in Hong Kong are justified as the means necessary for developing and expanding e-commerce.

In the case of China, emerging data privacy protection is likewise justified primarily by the need for such protection in order to develop e-commerce. The contrasts here with U.S. and German rationale are especially clear. There seems to be general agreement that historically, *individual* privacy was not a concern in China—both because of a long (especially Confucian) tradition of emphasizing the good of the larger society collectively, along with long practice of strong governmental control.

This historically *negative* posture towards the individual and individual privacy can be seen in the recent development of a new term for privacy in Chinese. According to Professor Lü:

> In 1985, the Chinese law dictionary only explained the concept of Yinsi. At that time, Yinsi[33] meant "privacy." But now people usually distinguish the difference between a "shameful secret" (Yinsi) and "privacy" (**Yinsi**) *[the two terms are pronounced the same but spelled differently in Chinese – editor]*. The term Yinsi is defined as a hidden, bad thing, while the term **Yinsi** is defined as personal thing people do not wish to tell others or to disclose in public. (Lü 2005)

Given this, perhaps it is fair to say that the meaning of Yinsi in 1985 was the

---

[33] According to Professor Zhang, this term has "no equivalent in Japanese but this term means something shameful or something wrong in a private and hidden area of life." (Zhang 2005)

closest equivalent to "privacy" in Chinese—but one that had overtones of bad, hidden things; subsequently, a relatively new (more positive, and thus closer to the Western) notion of privacy—**Yinsi**—has begun to emerge.[34]

Again, Chinese citizens enjoy some rights to data privacy protection—if comparatively limited as compared to those expected and enforced in many Western countries. Again, as with Hong Kong, the rationale for such protection is primarily as a means towards commerce and business efficiency (Chen 2005).

This overview of privacy expectations and data privacy protection laws suggests in a final way the possibility of a pluralism that holds together a shared value alongside its different manifestations in diverse cultural settings. In this case, the shared values of privacy and data privacy manifest themselves as protected under the laws of the U.S., Germany, Hong Kong and China: but these values are *interpreted* and applied in different ways. Simply, the rights to privacy and data privacy protection in China and Hong Kong are somewhat limited when compared with their counterparts in the U.S. and Germany; and they are justified primarily by appeals to economic interests rather than to notions of the individual and citizen required for democratic governance. But the appeal to economic interests is shared with the German law—and both Germany and the U.S. share similar notions of privacy as critical to the individual and democracy. In the end, despite the clear differences between, say, China on the one hand, and the U.S. on the other, regarding rationale and implementation—a recognizable notion and practice of privacy and data privacy protection is to be found in both countries, one that is shaped by and thus preserves their distinctive cultures, histories, and traditions.

---

[34] This initially negative sense of individual privacy thus parallels the Buddhist emphasis on *musi* (no self) we have seen in Japan. By the same token, "privacy" appears primarily as a loan-word in Japanese (Mizutani et al 2004, 121). Similarly, Priongrong Ramasoota notes that "The Thai language does not have a word for privacy but refers to it by descriptively translating from English as *khwam pen suan tua* or *khwam pen yu suan tua*, meaning "the state of being private" (2001, 97)." Ramasoota goes on to argue that "the traditional Thai conception of privacy is fundamentally *collectivistic*," i.e., one shared in the first instance "by *intimate members of the same household*. By this token, individualistic privacy is said to have no place in traditional Thai culture" (98). And as we saw in the Japanese example, Western-style notions of the individual and privacy directly *conflict* with largely negative attitudes in Buddhism towards notions of the individual or self:

> Buddhist thinking sees the obsession with one's individual self and one's possession, material or not, as the root source of suffering. Emancipation … means disillusionment with and relinquishing of selfdom and worldly desires. Therefore, individuality can be seen as both the beginning and the end to human emancipation in Buddhism. (Ramasoota 2001, 100f.)

(I am indebted to Prof. Soraj Hongladarom for making me aware of this article.)

Concluding Remarks: *Pros Hen* Equivocals and Pluralism

As Aristotle said of Being, "Privacy is said in many ways": not surprisingly, we find that there are different (and *changing*) meanings of "privacy" in U.S., E.U., Japan, China, Hong Kong, and Thailand. So, as we saw in Western examples of pluralism, shared focus on *expectations,* coupled with diverse cultural and national expectations regarding informed consent for recording in public spaces (Norway / U.S.), and shared assumption of "First Principles" of Data Protection, alongside diverse cultural / national approaches to implementing such protection (U.S. / E.U.)—so we see a similar pluralism operate *between* Western and Eastern traditions, cultures, and countries, in their approach to privacy and data privacy protection that show both *divergence* in concepts, justifications, and extant laws—reflecting the distinctive cultural / historical / political traditions and values of these very diverse countries, but in ways that likewise entail *convergence / resonance / harmony* [*he*] between basic elements and practices of Information Ethics.

To be sure, the needed cross-cultural dialogues regarding privacy, data privacy protection, and larger issues in Information Ethics are only at their beginning stages and much in this initial sketch will likely prove to be terribly incomplete, if not simply mistaken. As well, as the example we have seen of Internet research in Japan makes clear, not all differences can be smoothly reduced to a structure of pluralism: rather, real differences will remain and intransigent conflicts will also occur.[35] But, if I'm at least partly correct here, then it appears that we can discern structures of *ethical pluralism* emerging between East and West with regard to such basic concepts and issues as privacy and data privacy—a structure that provides us with both (quasi)-universal ethical norms while articulating and respecting the distinctive values, traditions, and ethical approaches that define national and cultural identities. Such pluralisms, I suggest, may make a useful contribution to the task of developing at least quasi-universal ethical norms in Information and Computer Ethics for a world whose often radically diverse cultures and peoples are increasingly wired together through computer-mediated communication technologies such as the Internet and the World Wide Web.

References

Ames, Roger T. and Henry Rosemont, Jr. 1998. *The Analects of Confucius: A Philosophical Translation*. New York: Ballantine Books.

AoIR (Association of Internet Researchers). 2002. Ethical Guidelines for

[35] See also note 3, above.

Internet Research. www.aoir.org/reports/ethics.pdf.

Burkhardt, Jeffrey, Paul. B. Thompson, and Tarla Rae Peterson. 2002. The first European congress on agricultural and food ethics and follow-up workshop on ethics and food biotechnology: a US perspective. *Agriculture and Human Values* 17.4: 327-332.

Bizer, Johann. 2003. Grundrechte im Netz: Von der freien Meinungsäußerung bis zum Recht auf Eigentum. In Schulzki-Haddouti Christiane (Hrsg.), *Bürgerrechte im Netz,* 21-29. Bonn: Bundeszentrale für politische Bildung. http://www.bpb.de/publikationen/UZX6DW,,0,B%fcrgerrechte_im_Netz.html.

Chen, Yang. 2005. Privacy in China. Unpublished Master's thesis. Nanyang Technological University, Singapore.

Code of Federal Regulations Title 45, Department of Health and Human Services, National Institutes of Health, Office For Protection From Research Risks, Part 46, Protection Of Human Subjects: http://www.hhs.gov/ohrp/humansubjects/guidance/45cfr46.htm.

Elgesem, Dag. 2002. What is Special about the Ethical Issues in Online Research? *Ethics and Information Technology* 4.3: 195-203. http://www.nyu.edu/projects/nissenbaum/ethics_elgesem.html.

Ess, Charles. 2003. The Cathedral or the Bazaar? The AoIR document on Internet Research Ethics as an Exercise in Open Source Ethics. In *Internet Research Annual Volume 1: Selected Papers from the Association of Internet Researchers Conferences 2000-2002*, ed. Mia Consolvo, 95-103. New York: Peter Lang, 2003.

—. 2004a. Beyond *Contemptus Mundi* and Cartesian Dualism: Western Resurrection of the BodySubject and (re)New(ed) Coherencies with Eastern Approaches to Life/Death. In *Philosophie des Todes: Death Philosophy East and West*, eds. Günter Wohlfart and Hans Georg-Moeller, 15-36. Munich: Chora Verlag.

—. 2004b. Moral Imperatives for Life in an Intercultural Global Village. In *The Internet and Our Moral Lives*, ed. Robert Cavalier, 161-193. Albany, New York: State University of New York Press.

—. 2005a. Discourse Ethics, in *Encyclopedia of Science, Technology, and Ethics,* ed. Carl Mitcham et al. New York: MacMillan Reference.

—. 2005b. Can the Local Reshape the Global? Ethical Imperatives for Humane Intercultural Communication Online. In *Localizing the Internet. Ethical Aspects in an Intercultural Perspective*, eds. Johannes Frühbauer, Rafael Capurro and Thomas Hausmanninger. Volume 4, ICIE Series.

—. 2005c. What is Information Ethics? http://www.drury.edu/ess/CAP04/cap04infoethics.html.

—. Forthcoming. Culture and Communication on Global Networks: Cultural

Diversity, Moral Relativism, and Hope for a Global Ethic? Or: Can Plato's *cybernetes* / Confucius" *junzi* navigate an interconnected world? In *Moral Philosophy and Information Technology*, eds. John Weckert and Jeroen van den Hoven. Cambridge: Cambridge University Press.

Johnson, Deborah. 2001. Privacy (ch. 5). In *Computer Ethics*, 109-136. 3rd. ed. Upper Saddle River, New Jersey: Prentice-Hall.

Lü, Yao-huai. 2005. Personal email to the author.

Mizutani, Masahiko, James Dorsey and James H. Moor. 2004. The internet and Japanese conception of privacy, *Ethics and Information Technology* 6: 121–128.

Nakada, Makoto. 2005. Personal email to the author.

Nakada, M., T. Tamura, L. Tkach-Kawasaki, and T. Iitaka. 2004. Does Old Japan Determine New Japan? The Relationship between *Seken*, the Internet, and political consciousness in Japan. In *Proceedings Cultural Attitudes Towards Communication and Technology 2004,* eds. F. Sudweeks and C. Ess, 143-156. Murdoch, Western Australia: Murdoch University. Australia,

NESH (National Committee for Research Ethics in the Social sciences and the Humanities). 2001. Guidelines for research ethics in the social sciences, law and the humanities. http://www.etikkom.no/NESH/guidelines.htm.

—. 2003. Research ethics guidelines for internet research: http://www.etikkom.no/Engelsk/Publications/internet03/view.

Persondataloven [Personal Data Law, Denmark]. Lov. Nr. 429 af 31. Maj 2000 som ændret ved lov. Nr. 280 af 25. April 2001: http://147.29.40.90/_GETDOC_/ACCN/A20000042930-REGL. English version, http://www.datatilsynet.dk/eng/index.html. (accessed 2 December 2003).

Ramasoota, Pirongrong. 2001. Privacy: A Philosophical sketch and a Search for a Thai Perception. *MANUSYA: Journal of Humanities* 4.2: 89-107.

Reidenberg, Joel R. 2000. Resolving Conflicting International Data Privacy Rules in Cyberspace. *Stanford Law Review* 52: 1315-1376.

Stahl, Bernd Carsten. 2004. *Responsible Management of Information Systems.* Hershey, PA: Idea Group Publishing.

Tamura, Takanori. 2004. Unpublished MS.

Thomson, Judith Jarvis. 1971. A Defense of Abortion. *Philosophy and Public Affairs* 1: 47-66.

Tu, Wei-Ming. 1999. Humanity as Embodied Love: Exploring Filial Piety as a Global Ethical Perspective. In *Jen Agape Tao with Tu Wei-Ming*, eds. Marko Zlomislic and David Goicoechea, 28-37. Binghampton, NY: Institute of Global Cultural Studies, 1999.

Walther, Joe. 2002. Research Ethics in Internet-Enabled Research: Human Subjects Issues and Methodological Myopia. *Ethics and Information*

Technology, 4: http://www.nyu.edu/projects/nissenbaum/ethics_walther.html.

Yasukawa, Hajime, and Ando Taro. 2002. Online Therapy—Characteristics and Problems for Media Mediated Social Service Activities. http://ofc-hjm.misc.hit-u.ac.jp/hjm/MyDesk/Bib/2002a_1.pdf.

Zhang, Ning. 2005. Personal email to the author.

Some specific resources for China

General

Bu, Wei. Forthcoming. Access to and Use of the Internet in China. In *Technology in a Multicultural and Global Society*, eds. May Thorseth and Charles Ess. Trondheim, Norway: NTNU Publication Series.

Fang, Weigui. 2004. *Das Internet und China: Digital sein, digitales Sein im Reich der Mitte*. Hannover, Germany: Heise.
(Many of these chapters began as articles for the German online magazine Telepolis, and so can be found there as well.)

Lagervist, Johan (guest editor). 2003. Chinese Intellectuals' Thoughts on the Internet. Special issue of *Contemporary Chinese Thought* 32. 2.

Lü, Yaohuai. 2002. Xin Xi Lun Li Xue *(Chinese Information Ethics)*. Middle South University Publication, Hunan.

Randolph Kluver (Singapore Internet Research Centre, School of Communication and Information, Nanyang Technological University) has developed an extensive bibliography on China and the Internet. Available from the author: <TRKluver@ntu.edu.sg> See also: http://www.ntu.edu.sg/sci/sirc.

Research Ethics

Döring, Ole. 2003. China's struggle for practical regulations in medical ethics. *Nature Reviews | Genetics* 4: 233-239. (PDF available online)
Excellent overview of efforts to develop Western-style medical research ethics in China, vis-à-vis both traditional cultural orientations (Confucianism first of all) and more recent economic, social, and political developments as China, especially through its membership in the WTO, must move towards more greater harmony with Western codes and practices.

Sang, Guowei. (n.d.) Ethical Principles and Protection of Trial Subjects in China. China Center for Pharmaceutical International Exchange. http://ccpie.haoyisheng.com/narticle2.php?id=429.
Provides an overview of the development of human subjects protections

practices in China's "Good Clinical Practice" (GCP) implementation—as defined first of all in the Helsinki declarations.

Privacy issues

Yang, Chen. 2005. Privacy in China. Unpublished Master's thesis. Nanyang Technological University, Singapore

Tang, Raymond. 2002. Approaches to Privacy ~ The Hong Kong Experience. http://www.pco.org.hk/english/infocentre/speech_20020222.html.

McDougall, Bonnie S. 2001. Briefing paper: concepts of privacy in English. http://www.arts.ed.ac.uk/asianstudies/privacyproject/briefing.html.

McDougall, Bonnie S. and Anders Hansson, eds. 2002. *Chinese Concepts of Privacy*. Leiden: Brill.
See also: The Privacy Project (University of Edinburgh, Asian Studies):
http://www.arts.ed.ac.uk/asianstudies/privacyproject/index.html.

Chiu, Michelle, Anthony Lee and Tsui Hou Ming. N.d. Protection of privacy in China, Hong Kong and Taiwan.
http://newmedia.cityu.edu.hk/cyberlaw/gp15/cnlaw1.html.
(This is from an online course on "Protection of Privacy in China, Hong Kong and Taiwan," created by Michelle Chiu, Anthony Lee and Tsui Hou Ming: very good overview!)
A more complete account of privacy rights and law in the PRC is available at http://www.privacyinternational.org/survey/phr2003/countries/china.htm.

Caslon Analytics. Privacy Guide. (Overview of Asia):
http://www.caslon.com.au/privacyguide6.htm#landscapes.

Tsui, Lokman. 2003. Online Privacy in China—A Preliminary Exploration. Unpublished manuscript. Available from the author:
<lokman.tsui@gmail.com>

CHAPTER SIX

A BUDDHIST MODEL FOR THE INFORMATIONAL PERSON

KEN HEROLD
HAMILTON COLLEGE

Prologue

Early proponents of the information age characterized the new world of intelligent machinery as a territory ripe for exploration and exploitation, as an electronic frontier in cyberspace. The ecological aspect of this metaphor was especially unfortunate given humankind's iniquitous record of despoiling geographic frontiers. Nevertheless, the general sentiment of this period of automation accentuated a revolutionary phase of knowledge discovery preceding and outstripping the more mundane, routine, and hoped-for successful settlement of the virtual spaces of the new privileged electronic communities. Reflecting both the prodigious rapidity of these developments and their unusually hidden and arguably deceptive powers, antagonists of the view variously attacked the computer, its technology, its science, and sometimes all of science. Clear efficiencies in the processing of vast data stores and transaction and communications streams, plus the general progress achieved in formulating systems with which to order and model complex events has muted the intensity of some fatalistic assessments of information at the societal level.

Most recently, however, the perceived nature of information itself has been criticized at a personal level. Borgmann (1999), for example, has posited information as directly usurping reality as opposed to providing us a basis for or about reality. Similarly, and from a Buddhist perspective, Hershock (1999) characterizes the negative effects of the contemporary information society on its members as a "colonization of consciousness." Introna (2001) argues that this electronic mediation, if considered to be fundamentally arbitrary per Baudrillard, leads to an impossibility of ethics in the information age. Cyberspace becomes a network of "hyperreal effects and representations that come *before* us, but do not *involve* us." [emphasis in original] Introna sees electronic mediation as the destroyer of obligations built through face-to-face experience of each other. The reified information entity dictates the contents of our mental life. Thus, another Buddhist author, Laycock (1994), poses the existence of a kind of digital *samsara*:

> When the world is attended primarily through electronic and print intermediaries, it is quite naturally taken to be given, not something underdetermined and limitlessly responsive, but a thing that is. Moreover, it is a world that mediates or stands between each one of us and all others even as it connects us.

To the defense of the informational person, Floridi (1999) raises a uniquely responsible and ethically-challenging position, posed more succinctly in reference to the history of Western thought, asserting the emergence of a transpersonal information capability filled with potential devotion and service to humankind:

> … an artificial universe which becomes, to an ever-increasing degree, the very environment we inhabit, and whose challenges we must answer. It is a domain as completely autonomous from each of us, as individual minds, as the physical world is, and which, unlike the latter, is capable of infinite growth in extension and complexity….this new environment has been created by the human mind, and since we are totally responsible for its existence, we may also be confident of its potential full intelligibility….at least since the Renaissance the mind has constantly moved away from reality, intentionally constructing and refining its own cultural environment as an alternative to the world of nature.

The paper explores a metaphysics of information enriched by a computational view of Buddhism consistent with onto-ethics. To the extent that Floridi has explained the new philosophy of information as borrowing methods from computer science to approach philosophical problems computationally, I believe an applied philosophy of information can return the fruits of these results back to grounding issues in the practices of information technology. With this process we also foster a cross-fertilisation between Eastern and Western philosophies, in the larger, intercultural arena.

Part I

In one traditional Buddhist analysis, the authority of teachings has its conceptual source in the teacher, or guru-equivalent. Consider the four dependencies, cited by the 14th Dalai Lama (1982), from the *Mahāyāna-sūtra-alankāra-kārikā:*

> *of the teacher*: depend not only on the teacher's person, but on what he teaches.
> *of the teaching*: depend not on merely the sweetness of the words and so forth, but on the meaning.
> *of the meaning*: depend not on the interpretive meaning, which must be understood in a different way, being controvertible since it was taught with a specific intent [or] under a specific circumstance, but depend upon the definitive meaning.
> *of the definitive meaning*: depend not on a consciousness with dualistic views, but on non-conceptual direct knowledge focused on [sunyata]

In the contemporary situation we are confronted with an ambiguity of information resources. Communications technology makes us more aware than ever of our dependence and reliance on an ordered approach to reliable information. Suppose we make use of the traditional analysis to restate the parameters by which authoritativeness of information resources should be investigated. This suggests a cycle of considerations from the person as authority, an external contribution, to levels of meaning contingent upon a perfection of the mind, an internal attribution. Suppose further there exists a new, informational metaphysics by which we may elabourate the person as information resource and the personal understanding of information. One approach is to study data and algorithm (instruction) as foundational entities of investigation of semantics and learning. Such is the derivation of information ethics (IE) from computer ethics as elabourated by Floridi and others.

In a pre-foundational exposition of IE, Moor (1985) describes a crucial ingredient for the field of computer ethics: the ubiquitous applicability of general-purpose manipulations of logic itself he terms "logical malleability."

> Logical malleability has both a syntactic and a semantic dimension. Syntactically, the logic of computers is malleable in terms of the number and variety of possible states and operations. Semantically, the logic of computers is malleable in that the states of the computer can be taken to represent anything.

This sense of ubiquity, where information is supposed to be capable of taking *any form*, discloses the ramifications of possible worlds in an onto-ethics (Cheng 2004) of information. First, that information may also suit *any purpose*, suggesting a universality which may be either fully intelligible or else wholly threatening depending on its reception. Second, information may acquire in its process of formation an *enhanced status* and pseudo-objectivity. Third, despite this initial mutability, information *may not change*, that is, in the process of manufacturing, a fixity occurs which may not be expected or welcome. The metaphysics of these transformations are subjects of investigation under the rubric of the philosophy of the information "life-cycle."

At the inception of the computer age, Turing realized that the pulses emanating from his so-called intelligent machinery could not be differentiated *a priori* between instances of content (data) or instruction (meta-data). If the mechanization of thought processes or techniques specifically led to such a conclusion, then, if generalized, this means that information cannot be reduced—it may only be transformed. That is, another metaphysical basis of the philosophy of information is this ever-potential equivalency, evident as well in terms of the complementarity of structure and function. Thus it is not always clear, when one is dealing with cybernetic inputs and outputs at a certain level of abstraction, what is formally constructing or constructed to the mind.

One long generation of software development has evidenced value judgments implemented within a computer programming environment that is exceedingly complex and only partially knowable even to the programmers themselves. Moor refers to this third complicating factor as "invisibility." If there existed an underlying digital plenum, then this invisibility would obscure its compositional action and its transformational patterns in information objects would be unclear or seem unsupported. A very young Norbert Wiener (1914) considered such partial knowledge systems from an individual compositional standpoint, questioning the persistence of the self and the nature of identity over time. Wiener understood the infinite complexity of experience to depend upon the lack of self-sufficiency of information objects. He foresaw our right to a formalized doubt for *every* concept, without the duty to exhaustively examine *any* concept; in short, a liberating and not an enslaving relativism. Furthermore, the character of the moment of experience itself, though given as a system possessing coherence in cross-section of time, is also insufficient. In Buddhist terms, this lack of self-sufficiency and system of dependent relations is the *śūnyatā* (emptiness) of definitive meaning and the dependent-arising of all phenomena. This type of complexity is a fundamental type of informational complexity of the forms of experience. Following Collier (1999) in his adaptation of Bennett's notion of logical depth, it is fundamental "in the sense that anything that is complex in any other way must also be informationally complex."

Within this onto-ethics of information, the class of informational entities includes both persons and non-persons as authoritative resources. Semantic logical malleability, together with description-instruction complementarity, could explain our attribution of authoritative personhood to digital entities, as well as our interpretation of real persons in terms of their informational surrogates. Computational Buddhism also considers the origination and source of control for information entities, the nature of their physicality, their interactivity as momentariness, and the possibility of an informational realism or informational emergentism within an evolved and liberating infosphere. The remainder of this section is a sketch of each of these aspects in brief drawn from the field of computing. Key to this outline is extending Wiener's notion of the system-moment as an informational structure into the realm of our essential experience of software. An interesting neurological correlate may be found in Brown (1999). The sketches are admittedly terse and incomplete and mostly serve as the context for future elabouration.

Control is the defining notion of cybernetics and the instrument of information systems engineering. Simon (see Cordeschi 2002) played a key role in equating the person as human organism with the computer program, together as examples of an information processing system for problem solving. Floridi's LoA's (Floridi et al, 2004) and other models simulating knowledge processes, such as Lynch (1997) and Sloman (2000), exhibit diagrammatic schemas in which feedback loops channel complex messages. These messages may be described as information objects themselves, per our complementarity notion above. These are not unique at all within computation but offer constructive models for testing against real situations, as simulations or isomorphisms. These control models also share a characteristic time element: a non-linear, branching, and structured temporality of the computational present, one not limited to physical time. In information metaphysics the most important meaning of control is that of the experimental method: an authentic, identifiable versioning of data, as communicated in meta-data, for testing and verification purposes in systems. In interactive and necessarily participatory systems, this requires a continual openness and making explicit and accessible all underlying assumptions and design decisions. Collier's account of algorithmic information theoretic causation suggests a link between a dynamical account of Bennett's logical depth and this computational view of time. Thus, control is of persistence and the change of persistence is the creation of meaningful contexts.

The quantum theory of computation has advanced the idea that information is physical. It seems deserving in turn to consider physicality as informational: as the abstract features of particular instances. Goodman (2004) links the Western concept of physical tropes with Vasubandhu's notion of *dharmas,* citing earlier works by Ganeri and by Siderits. In my reading of Goodman I would not choose between alternatives in his philological analysis of *dharmas* as either properties or events, but instead regard information, or something like an informational trope, to be both. Information certainly has qualities of both process and thing. The vague boundary between the entities of hardware and software is an instructive case. Another example might be the physical aspect of memory in molecules. Meirav's (2000) criterion for physicality (that a physical object corresponds in a strict mereological structure in relation to the spatial domain it occupies) could be modified towards informationality as a relation for all objects, per Collier. Physicality as informational entails a reappraisal of artificiality in the context of controlled systems of feedback. Cordeschi (2002) propounds five theses for the artificial which may be applied to further discovery of the metaphysics of the system-moment artifact: functional, modelling, representational, mentalist, and explanatory.

Whitehead's notion of actual occasion acquires fresh meaning when interpreted through computational means as a structured present moment, as if Wiener's system-moments were a software model. Its constitution, or concrescence, unites the person with emerging experiences in computational time and the person constructs value or education through the positive selection of data. Such knowledge events are momentary, cumulative, and creative. Whitehead refers to "drops of experience" both as constructed and constructing, transcending the subject-object distinction. Education in this Whiteheadian model is in essence interactive and occurs under an undefined, particular mode of "wisdom." The Buddhist correlate is a non-dual wisdom consciousness. The cross-sectional coherence identified by Wiener as persistence of personal identity can be thought of as a network of information objects messaging each other in the infosphere. Possible explanations for informational access and presence with respect to these computational drops (or windows) of experience are propounded by Morin (2004) in his theory integrating inner speech with problem-solving "where the self represents the problem and self-information the solution." In this sense, some Buddhist meditation, and in our analogy software simulation, may be seen as problem-solving sessions in which inner speech conducts a critical rational analysis of reality through the manipulation of information about information.

Floridi's (2004b) informational realism (IR) has its precedent in Bateson and Spencer-Brown, what can be termed a "distinction ontology." Information is all that there is but it is not a substance, it is a relation. IR undermines the usual implications of a substance ontology towards change, in the Western view, and rejects ontological absolutism and commitment to first-order statements outright, like the Buddhists. I suggest that his level of abstraction (LoA) can be viewed as a distinguishable trope. Tropes could be information objects themselves, as dharmas. IR may comport with Vacaspatimisra that "attributes are not something apart from the substance of a thing, but productive imagination constructs them as something different." (Puhakka 1975) From Floridi and Sanders (2004):

The LoA is determined by the way in which one chooses to describe, analyse and discuss a system and its context. The "Method of Abstraction" is explained in terms of an "interface" or set of features or observables at a given "LoA". Agenthood, and in particular moral agenthood, depends on a LoA. Our guidelines for agenthood are: interactivity (response to stimulus by change of state), autonomy (ability to change state without stimulus) and adaptability (ability to change the "transition rules" by which state is changed) at a given

LoA. Morality may be thought of as a "threshold" defined on the observables in the interface determining the LoA under consideration. An agent is morally good if its actions all respect that threshold; and it is morally evil if some action violates it. That view is particularly informative when the agent constitutes a software or digital system, and the observables are numerical.

Tropes are not an uncontroversial philosophical device invented to address problems of compositionality and continuity of identity. The potential affinity of tropes with dharmas has been advanced by Goodman with respect to the momentariness of existents. Dharmas, the fundamental analytic unit of things both physical and mental, precede Abhidharma theory and have two major elabourations with respect to inherent existence. Mādhyamaka theory treats dharmas themselves as fully relational, lacking inherent existence or causal power. Thus all phenomena are composite and momentary identities whose continuity or persistence is real, though illusory. That is, causality is accomplished through networks of relations, through dependent origination. The lack of intrinsic nature, known as emptiness or openness, does not deny causal regularity but itself depends upon it. Perrett (2004) borrows a logical argument structure from the 11th century Ratnakirti, the Master Argument for Momentariness: any simple possessed of intrinsic causal powers must be a momentary entity. This represents the alternative view to Madhyamaka, that dharmas possess intrinsic causal power.

Other possible computational interpretations of trope theory parallel its disparate branches. One might study apparent immanent universals in Buddhist metaphysics and the notion of a single universal instantiated within multiple particulars, per Armstrong. Or a trope-based unity of particulars could be explored as an alternative following Dharmakirti (Mortensen 2004). Perhaps there are precursors to an informational emergentism in the bundle version of tropes of Campbell, in which basic individuals are mereological bundles of compresent (at least spatiotemporally co-located) tropes. Multiple tropes are always compresent in a given location, or alternately, every trope type requires some requisite number of other compresent tropes. In Simons' nuclear-theory alternative for tropes, a requisite core of compresent tropes may be more essential as kinds instanced together per basic causal laws defining an enduring individual. In Martin's substance-attribute version, tropes, even bundles of tropes, are ontologically dependent entities, accounting for "thisness" or "thinginess" as they are, rather than in their own right. An informational trope theory may associate these variations with differing schools of Buddhist logic, the subject of future work.

In conclusion, the above framework is tentative and meant to indicate a potential computational view of Buddhism. Some features of Buddhist analytical meditation may share characteristics with the simulations and modelling presented in an applied philosophy of information. What portion of *upāya* is informational? The gist of the remainder of the paper observes Schroeder's (2000) finding of pedagogy as metapraxis.

Part II

The world's earliest dated printed book is a copy of the Diamond Sutra *(Jin gang ban ruo bo luo mi jing)* from 868 CE. In itself this may not be remarkable, although the combination of Buddhist subject matter with pioneering information technology of the day is evident. An extensive and comprehensive wisdom literature is bound to attract inventive forms of expression in accordance with its importance for civilizations. Confucian written scholarship and Christian typographic scripture are further examples. In the context of the transmission of knowledge, Varela (1999) says "the rediscovery of Asian philosophy, particularly of the Buddhist tradition, is a second renaissance in the cultural history of the West." More specifically, Wood (2004) identifies four key features of a general Buddhist influence on scholarship: an informed self-creation, an openness and inclusiveness based on compassion, a critical mindfulness, and an attentiveness to others. A Buddhist model for the informational person entails these features within the framework of an intellectual order seeking systematic compassion and an end to the suffering represented by its disorder. The cybernetic ethics of Wiener and information ethics of Floridi operate within this sphere of participatory computing, highlighted by the icon of the message becoming the interactive program. Whether or not personal erudition and cultural education are relatively well-understood in the age of printing; how does the informational model aid us in the practice of building a digital adaptation?

An historical examination of the emergence of this paradigm may be found in Herold's (2003) information continuum conjecture: library information structures and objects share this characteristic intellectual ordering of abstract human works. Library theory can guide information science and technology in the pursuit of new, digital scholarly systems. Consider the traditional path from author to reader, where the author's message is produced by a publisher and later digested, indexed, classified and categorized by a library as a kind of social annotation to the process. The library's effort conserves the original intellectual entity, the conversation of author with reader, and enhances the openness, inclusiveness, awareness, and accessibility of the original work to others, for the work as process. Similarly, Santaella-Braga (2004) proposes a transformation of dialogism into heterologism and the externalization of collective negotiations of meaning in the communitarian exchange of cyberspace objects. In this type of informational model, software comes to be seen as the external manifestation of inner speech and all of its ramifications. The inner speech of scholarly communication is like a transaction that takes place between individuals, self and others; this interface is brought to a new dimension of intensity in the digital network of relationships.

Derivation of an informational association between the wisdom literature produced by liberated minds and the learning processes and objects for minds seeking liberation comes from Fenner (1995). His systems-cybernetics model of Buddhist Middle Path analysis is based upon what he terms perspectivism and contextualization. Perspectivism acknowledges the contingency of conceptual frameworks attempting to structure these exemplars of natural systems of mental events. Contextualization refers to the influence of cultural, language and philosophical communities on the mediation and interpretation of cognitive and perceptual processes. What Fenner treats as exemplary, the discourse on the perfection of insight *(prajñā)* through the cognition of openness *(śūnyatā),* I suggest be considered as generalized throughout the larger metaphysical system of Abhidharma: instructional as simulation. As Fenner notes: "Meditational texts usually describe cognitive systems....This view, that textual materials describe systems of thought and matter, accords also with how Buddhist scholars understand their texts." Fenner's unique, therapeutic methodology echoes Floridi's early description of an informational model as a "network of logically interdependent but mutually irreducible concepts." Fenner seeks "a base-line model, in which the insight into openness is developed gradually over time and in dependence on specific causes and conditions." It is this "causal continuity" which forms a basis for Floridi's Method of Abstraction. Also, Fenner's strategy is reflected in the findings of Christman (2004) on narrativity, particularly when interpreted in unison with the notion of inner speech from Morin (2004) and Langsam's (2002) model of self-knowledge of thought contents. From Langsam:

> To be related to an object via an act of consciousness counts as having *access* to the object insofar as it enables one to acquire knowledge of the object. It is because consciousness is relational that we can be conscious of, and have conscious access to, the very external things [literally elements of the contents of our thoughts] we are consciously thinking about....we can have conscious access to external things both through perceptually experiencing them and through merely thinking of them. [emphasis in original]

In this section I explore the concept of *rūpa-citta vipra-yukta saṃskāra* in order to discover how it might contribute to an understanding of the information model, as suggested by Ames (private communication; 2003), and how information plays a role in the pursuit of wisdom.

The Abhidharma presents a detailed enumeration of ultimates such as mental factors, states of consciousness, and physical elements, among other things, as a logical analysis and exposition of fundamental principles governing the interactions between mind and matter. Some further justification for a Floridian study stems from his characterization of the philosophy of information as *liminal* or "not reducible to some form of externalism (naturalization of information) or internalism (information is in the mind of the beholder)" and *constructionist,* meaning we "neither discover nor invent the world, we design it." (Floridi 2004a) Although liminalist constructionism is not a tenet of belief about information itself, this balance between extreme views suggests a core of delicate debating points around which informational suppositions about *rūpa-citta vipra-yukta saṃskāra* may be poised. The Sanskrit phrase translates roughly as "conditioned factors that are neither matter nor mind." As a starting point for discussion, it surpasses Wiener's statement regarding the nature of information "Information is information, neither matter nor energy" and moves towards Sayre's (1976) proposal for Informational Realism as a bridging concept between the mental and the physical. Popper's World 3 is another potential domain of an informational realism.

Stcherbatsky (1962) describes the importance of this phrase as "forces which can neither be included among material nor among spiritual elements." A force is being interpreted here as "a condition, a fact, upon which another fact arises or becomes prominent by itself." Stcherbatsky's Indian Buddhist sources identify fourteen such forces, among them origination, subsistence, decay, extinction, the force imparting significance to words, the force imparting significance to sentences, the force imparting significance to articulate sounds. A full chart is found below. The ad hoc nature of these elements indicates an information metaphysics made pragmatic through the inclusion of a category of structures of this type. Why? These "forces" are utilized as fact-processing algorithms and are more like abstract tools than absolute states. Fenner reiterates a conceptual analysis or "reasoning" best classically described as Aristotelian predication for which one manipulates these internal discriminations and their elabourations. None of these are ultimates themselves, but conventional entities used experimentally to probe reality.

The 14th Dalai Lama (1966) gives a somewhat difficult, early rendering from the Tibetan context in his expression "heap of volitional formations containing mental-coefficients dissociated from consciousness (immaterial but not associated with);" the word "heap" is described as group, collection or aggregate *(skandha).* This is helpful in the sense of rendering the equivalent "forces" and "volitional formations" as a configuration of choices or options. A more "modern" translation from the Tibetan by Hopkins (1983) frames this alternately for us as "compositional factors which are neither form nor consciousness." The phrase "compositional factors" is quasi-mathematical and again suggests a model, schema, or template through which an analysis is conducted or a processing of facts takes place. In Zahler and Hopkins (1983) we find aggregates defined as "a type of skilful object " S.: *skandha*, T: *phung po*; form, S: *rūpa*, T: *gzugs*; compositional factors (limitless in number), S: *saṃskāra*, T: *du byed;* non-associated compositional factors, S: *viprayuktasaṃskāra*, T: *ldan min 'du byed.* They are called compositional factors, according to this authority, due to being factors that allow for the aggregation of causes and conditions and for the production, abiding, and cessation of products. "*Rūpa-citta vipra-yukta saṃskāra*" is of the nature of informational tropes. Like cognitive software (Rescher 2002), its computational aspect is abstract and dynamic, but at the same time particular and static.

The Dalai Lama explains why Buddhist treatises specifically classify all types of phenomena from the point of view of transforming the mind. He emphasizes the goal of impartially exploring the nature of objective reality in order to solve problems, and moreover "to gain a precise understanding of the objects apprehended by the mind. This leads one to investigate whether an object that appears to the mind actually exists in accordance with the way it appears" (Wallace 2003). Furthermore, the Dalai Lama classifies mental processes that facilitate the pursuit of wisdom as "skillful means," while those directly involved are called "wisdom." Under this rubric of the two truths, then, we might interpret *Rūpa-citta vipra-yukta saṃskāra* as conventional objects of an informational reality whose ultimate function is the generation of a liberating existence. Harvey (2000) distinguishes three senses for skilful means. The first refers to grouping the first five of the six Bodhisattva perfections under the idea of method, the sixth being wisdom; thus, the union of method and wisdom. The second refers to the adaptation of teachings or the modification of a teacher's manifestations to the level of an audience. The third type is the compassionate overriding of ethical precepts, a sometimes extreme adaptation of actions to circumstances. It is the second meaning which provides an objective for the employment of informational tropes: the transmission of teaching and method together. In his comprehensive treatment of skilful objects, Pye (1978) focuses closely on the persistence of Buddhist meaning: the enormously powerful adaptive quality of the teachings includes both the ability to construct provisional means of expression as well as the ability to destroy these when the path has been traversed or the mind transformed. This idea is not limited to Pye's formal topic of Mahāyāna concepts.

This interpretive schema posits two divisions of 24 factors:

Agenthood: Person/Individual/Sentience

Person	*Pudgala*	*gang zag*	so defined because of being designated in dependence upon a collection of form and consciousness. Since a person is neither form nor consciousness but impermanent, it can only be an instance of the remaining category of impermanent phenomena, a non-associated compositional factor

Other: Non-Person/Non-individual/Non-sentience

Non-person	*Apudgala*	*gang zag ma yin pa'I ldan min 'du byed*	These 23 factors are called "designations to states" viprayuktasa.mskaara

Acquisition	*Prāpti*	*Thob pa*	finding and possessing the increase and decrease of virtues and so forth
absorption without discrimina-tion	*asaṃjñasamāpatti*	*du shes med pa'I snyoms 'jug*	a lack of the coarse feelings and discriminations associated with the third concentration and below, produced in dependence on the fourth concentration by common beings only. Kelsang Gyatso explains this as a subtle state where the mind is single-pointedly absorbed in nothingness and suppressing gross objects and delusions from arising, yet not making progress towards eliminating subtle objects and delusions, p. 118.

absorption of cessation	*nirodhasamāpatti*	*gog pa'I snyoms 'jug*	achieved only by Superiors, in which there is a lack of the coarse feelings and discriminations associated with the peak of cyclic existence (the highest formless level) and below
one having no discrimina -tion	*āsaṃjñika*	*du shes med pa pa*	a person born among the gods in the condition of being without coarse feelings and discriminations
life faculty	*Jīvitendriya*	*srog gi dbang po*	designated to the state of living, the base of consciousness and warmth
similarity of type	*nikāyasabhāgata*	*rigs 'thun pa*	state or condition of likeness
Birth or Production	*Jāti*	*Skye ba*	Four states of characteristics of things:
Aging	*Jarā*	*Rga ba*	
Duration/ Non-duration	*Sthiti*	*Gnas pa*	
imperman ence	*Anityatā*	*Mi rtag pa*	
Group of stems	*Nāmakāya*	*Ming gi tshog*	Three states of verbal conventions; parsing?:
Group of words	*Padakāya*	*Tshig gi tshogs*	
Group of letters	*Vyañjanakāya*	*Yi ge'I tshogs*	
state of an ordinary being	*pṛthagjanatva*	*so so skye bo nyid*	one who has not attained the qualities of Superiors; Vaibhashikas call this "non-acquisition and do not count 9 final additional factors
Continuity	*pravṛtti*	*jug pa*	Non-interrupted state of a continuum of causes and effects

Distinction	*Pratiniyama*	*so sor nges pa*	in 3 types: a) particular/general, b) virtues-pleasures/sins-pains, c) causes/effects
relatedness	*Yoga*	*byor 'grel*	in 3 types: a) means (collection of tools), b) aggregation (reliance of causes with each other in a collection of causes), c) suitability (each thing's having its own function);
Rapidity	*Java*	*gyogs pa*	of the arising of effects immediately after their causes
Order	*Anukrama*	*go rim*	serial state of former and later, high and low, etc
Time	*Kāla*	*Dus*	states of the past, present and future
Area	*Deśa*	*Yul*	composite of a place and the persons therein; in Tibetan yul can mean a sphere or region, physical or metaphysical, e.g. cyberspace; infosphere
Number	*saṃkhyā*	*Grangs*	condition of measure; numerosity?
Collection	*Sāmagrī*	*tshogs pa*	complete collection of causes and specifically to that completeness; in Tibetan "tshogs pa" refers to a mass, group, multitude, or an accumulation of things

Recalling that "non-associated" in our present context means "not associated with mind or mental factors," there are also in this larger theory some 49 compositional factors associated with the mind, which together with feeling and discrimination account for 51 mental factors. Lopez (1987) The first 14 have in common with those identified by Stcherbatsky the distinction of being taught as part of the Abhidharma, or philosophical curriculum of Buddhist schools. A question deriving from an investigation of the concept of information within the wider Abhidharma would be the applicability of all abstract objects whose purpose in some philosophical traditions is to provide knowledge necessary for practicing meditators to attain a particular clarity and state of consciousness, when utilizing these factors as specific objects during special sessions of thinking known as analytical meditation. Lopez notes the mirage-like quality of discrimination, or one's distinction and identification of objects: "An individual's discriminations falsely appear to be self-sufficient possessions of the person due to considering one's own discriminations to be valid and those of others to be invalid."

With these non-associated factors Geshe Kelsang Gyatso (1993) gives the following outline of holdings by various schools: "the Vaibashikas believe that they find the "I" to be just the collection of the five aggregates. Some Sautrantikas believe the same as the Vaibashikas, whereas others believe the same as the Madhyamika-Svatantrikas, that the root mental consciousness is the "I." The Cittamatrins believe that the consciousness-basis-of-all is the "I." The Madhyamika-Prasangikas reject all these assertions. They realize that the "I" is merely imputed in dependence upon any of its aggregates....in reality all phenomena, including our "I," are merely imputed by mind and do not exist from their own side in the least" (p. 14). Thus, the Madhyamika-Prasangikas, or Middle Way-Consequentialists, pose an informational realism based upon imputation. A word about "self." The 14th Dalai Lama has described this in the understanding of the two truths according to the Yogacara-Cittamatrin view: "The existence of an object which is an entity external to the consciousness perceiving it is called "self." The absence of that self is called selflessness. And the emptiness of such existence is called emptiness. Therefore, they explain that all phenomena are included within the entity of the mind."

Finally, Goodman (2004) recounts Vasubandhu's classification of dharmas to include "non-mental, non-physical caused *dharmas* (*citta-viprayukta-sa skāra-s*)". Vasubandhu asserts an indefinite divisibility of objects, positing from the discovery of no partless phenomena upon investigation that phenomena are dependent-arising. Collier has offered us an interpretation of information as a vehicle for causation, bringing around our circle completely. In the Buddhist view, the object of cognition is nothing apart from the act of cognition, contrary to holding that "the identity of distinct objects—however fleeting and momentary—is considered to be prior to and the basis for the subsequent conceptual elabouration." (Puhakka 1975)

Macy's (1991) intuitive classic describes "mutual causality" within the context of general systems theory, hinting at a possible linkage between cybernetics and, as she puts it, the dharma of natural systems, perhaps dharmas. Griffin (1991) commends Macy for her discussions of the ethical implications of "the mutual co-arising of self and society, doer and deed, means and ends, personal transformation and social action, among other things." However, he points out that general systems theory does not possess the conceptual tools required for deeper probing of codependent origination, preferring in that regard a Whiteheadian-Hartshornean process hermeneutic instead. Griffin helpfully supplies six philosophical explications for "mutual causality" (in opposition to a unidirectional causality) stemming from his critique of Macy, worth summarizing here:

LINEAR
L1) order descends hierarchically
L2) single, isolable causes
L3) events are passive: effect pre-exists in cause
L4) realism, independent world (actualities)
L5) only internal relations (including non-actuals)

MUTUAL
M1) order is intrinsic and self-emanating
M2) multiple causation, not unipolar
M3) effects have emergent qualities
M4) mind-shaped reality
M5) internal and external relations

A certain reciprocity exists between structure and function in the construction of information objects, and so both structuralist and functionalist arguments will be discussed. Finally, a kinship is struck between arguments concerning the description of reality and the status of objects within the descriptive discourse.

Software repositories of knowledge are designed to be queried in an erotetic method through structured tools. The special system objects collabouratively created as information objects may be known as intellectual artifacts. These are artifacts in an informational sense in which the idea of "work" is preserved. These information objects are assumed to be inherently potential erotetic structures—skilful objects of a kind. Hongladarom (2004) interprets Borgmann as a criticism of the limitations of Floridi's (2003) information ethics (IE). Floridi (2004) answers the ethical question posed by IE "What is good for an informational entity and the infosphere in general?" The answer is provided by a minimalist theory of deserts: any informational entity is recognized to be the center of some basic ethical claims, which deserve recognition and should help to regulate the implementation of any information process involving it, if possible." Hongladarom directs our attention to a specific quality of membership in the infosphere, the needs of persons as informational entities with respect to a particular metaphysical aspect of information: its transparency, distinct from Moor's more negative notion of transparency. Positive transparency is the ready availability of "information about information" throughout the infosphere, a kind of service which advances the informational health of its members by providing them the means to transform and share local information globally and vice versa. "Floridi's infosphere should be populated only by those pieces of information that are well represented. That is, the pieces should be arranged in such a way that the mind finds it easy to work on them, such as through the cataloguing system used by libraries (but obviously this is not the only way)."[36] Cheng's onto-ethics reaches the same conclusion as Hongladarom as to the critical importance of persons understanding their intellectual environment. If computer ethics is indeed a kind of onto-ethics, then perhaps these rich legacy discourses into the common attributes of persons and non-persons will bear further investigation.

References

Ames, William L. 2003. Emptiness and quantum theory. In *Buddhism and Science: Breaking New Ground,* ed. B. Alan Wallace. New York: Columbia University Press.

Borgmann, Albert. 1999. *Holding On to Reality: The Nature of Information at the Turn of the Millenium* Chicago: University of Chicago Press.

Brody, R. 2003. Information Ethics in the Design and Use of Metadata. *Technology and Society Magazine, IEEE* 22(2) (Summer 2003) 34-39.

Brown, Jason W. 1999. Microgenesis and Buddhism: the concept of momentariness. *Philosophy East & West* 49(3): 261-277.

Cheng, Chung-Ying. 2004. Dimensions of the *Dao* and onto-ethics in light of the DDJ*. *Journal of Chinese Philosophy* 31(2): 143-182.

Christman, John. 2004. Narrative Unity as a Condition for Personhood. *Metaphilosophy* 35(5): 695-713.

Cordeschi, Roberto. 2002. *The Discovery of the Artificial: Behaviour, Mind and Machines Before and Beyond Cybernetics* Dordrecht: Kluwer.

Cox, Collett. 1995. *Disputed Dharmas: Early Buddhist Theories on Existence; An Annotated Translation of the Section on Factors Dissociated from Thought from Sanghabhadra's Nyayanusara.* Tokyo: International Institute for Buddhist Studies.

Fenner, Peter. 1995. *Reasoning Into Reality: A System-Cybernetic Model and Therapeutic Interpretation of Buddhist Middle Path Analysis.* Boston: Wisdom.

---

[36] Herold (2003) describes the historical and emerging importance of metadata as information about information in exactly this regard. Brody (2003) suggests the application of two of Severson's principles of information ethics, fair representation and non-malificience, and all that they imply, as "the foundation for [IEEE] engineers' ethical decisions in [metadata]" Her pioneering examination of metadata ethics covers the entire information life cycle as originally introduced by Floridi (1999).

Floridi, Luciano. 1999. *Philosophy and Computing: An Introduction.* London; New York: Routledge.

—. 2003. On the intrinsic value of information objects and the infosphere. *Ethics and Information Technology* 4(4): 287-304.

—. 2004a. LIS as applied philosophy of information: a reappraisal. *Library Trends* 52(3): 658-665.

—. 2004b. Informational realism. In *Computers and Philosophy 2003—Selected Papers from the Computer and Philosophy conference (CAP 2003),* eds. J. Weckert. and Y. Al-Saggaf, 7-12. Conferences in Research and Practice in Information Technology, 37. ACS.

Floridi, Luciano and J. W. Sanders. 2004. On the morality of artificial agents. *Minds and Machines* 349-379.

Floridi, Luciano et al. 2004. The philosophy of information: a methodological point of view. http://www.wolfson.ox.ac.uk/~floridi/pdf/tpoiampov.pdf (accessed November 13, 2004).

Goodman, Charles. 2004. The *Treasury of Metaphysics* and the physical world. *The Philosophical Quarterly* 54: 389-401.

Griffin, David Ray. 1991. Review of *Mutual Causality in Buddhism and General Systems Theory: The Dharma of Natural Systems*, by Joanna Macy. *Process Studies* 20(4):244-248.

Gu, Ming Dong. 2003. The Taiji Diagram: A meta-sign in Chinese thought. *Journal of Chinese Philosophy* 30(2):195-218.

Gyatso, Tenzin, the Dalai Lama XIV. 1966. *The Opening of the Wisdom Eye*

—. 1982. A key to the Madhyamaka. In *Four Essential Buddhist Texts, 1st Revised ed.* Dharamsala: Library of Tibetan Works and Archives.

Gyatso, Tenzin, the Dalai Lama XIV. 2003. Understanding and rransforming the mind. In *Buddhism and Science: Breaking New Ground,* ed. B. Alan Wallace. New York: Columbia University Press.

Geshe Kelsang Gyatso. 1993. *Understanding the Mind: Lorig An Explanation of the Nature and Functions of the Mind.*

Hall, Patrick A. V. 1975. *Computational Structures: An Introduction to Non-numerical Computing* New York: Macdonald Jane's/American Elsevier.

Harvey, Peter. 2000. *An Introduction to Buddhist Ethics: Foundations, Values and Issues.* Cambridge: Cambridge University Press.

Hayward, Jeremy W. 1987. *Shifting Worlds, Changing Minds: Where the Sciences and Buddhism Meet.* Boston: Shambhala.

Herold, Ken. 2001. Librarianship and the philosophy of information. *Library Philosophy and Practice* 3(2).

—. 2003. An information continuum conjecture. *Minds and Machines* 13: 553-566.

Hershock, Peter D. 1999. *Reinventing the Wheel: A Buddhist Response to the*

Information Age Albany: SUNY Press.

Hongladarom, Soraj. 2004. Making information transparent as a means to close the global digital divide. *Minds and Machines* 14(1): 85-99.

Hopkins, Jeffrey. 1983. *Meditation on Emptiness*, citing as source: dKon-mchog-bstan-pa'I-sgron-me*, 1762-1823. *Beginnings of Annotations on (Dzong-ka-ba's) "Essence of the Good Explanations" on the Topic of Mind-Only, Illumination of a Hundred Mind-Only Texts.*
(*Also identified as Gungtang Konchok Tenpay Drunme; Konchog Tenpai Dronmey; Gung Tang; Gun-than Dkon-mchog-bstan-pa"I-sgron-me; Gung-thang-pa; rJe Gung thangpa; or as Gonchok Denbay Dronmay.)

Introna, Lucas D. 2002. The (im)possibility of ethics in the information age. *Information and Organization* 12: 71-84.

Langsam, Harold. 2002. Externalism, self-knowledge, and inner observation. *Australian Journal of Philosophy* 80: 42-61.

Laycock, Steven. 1994. *Mind as Mirror and the Mirroring of Mind: Buddhist Reflections on Western Phenomenology* Albany: SUNY Press.

Liu, Gang. 2004. Cyberphilosophy and a possible foundation for the future oriental philosophy of techno-science in a framework of metaphilosophical pluralism. http://www.blogchina.com, personal correspondence.

Lopez, Donald S. 1987. *A Study of Svātantrika.* Ithaca: Snow Lion.

Macy, Joanna. 1991. *Mutual Causality in Buddhism and General Systems Theory: The Dharma of Natural Systems.* Albany: State University of New York Press.

Lynch, M. 1997. Towards a paradigm of information system. *Philosophical Aspects of Information Systems*, eds. Probert Winder and Beeson, eds. London: Taylor & Francis.

Meirav, Ariel. 2000. A mereological criterion for physicality. *The Southern Journal of Philosophy* 38(4): 619-631.

Morin, Alain. 2004. Possible links between self-awareness and inner speech: theoretical background, underlying mechanisms, and empirical evidence. http://cogprints.ecs.soton.ac.uk/archive/00003784 (accessed September 6, 2004).

Moor, James H. 1985. What is Computer Ethics? *Metaphilosophy* 16(4): 266-275

Mortensen, Chris. 2004. "Dharmakirti and Priest on change. *Philosophy East & West* 54(1): 20-28

O'Connor, Timothy and Jonathan D. Jacobs. 2003. Emergent individuals. *The Philosophical Quarterly* 53: 213.

Perrett, Roy W. 2004. The momentariness of simples. *Philosophy* 79: 435-445.

Pickering, John. 1995. Buddhism and cognitivism: a postmodern appraisal. *Asian Philosophy* 5(1): 23.

Puhakka, Kaisa. 1975. *Knowledge and Reality: A Comparative Study of Quine and Some Buddhist Logicians.* Delhi: Motilal Banarsidass.
Pye, Michael. 1978. *Skilful Means: A Concept in Mahayana Buddhism* London: Duckworth.
Rescher, Nicholas. 2002. Process philosophy. In *The Stanford Encyclopedia of Philosophy (Summer 2002 Edition)*, ed. Edward N. Zalta. http://plato.stanford.edu/archives/sum2002/entries/process-philosophy/ (accessed November 20, 2004.
Santaella-Braga, Lucia. 2004. Interactivity in the light of dialogism. *Semiotica* 148-1/4: 119-135.
Sayre, K. 1976. *Cybernetics and the Philosophy of Mind.* London: Routledge & Kegan Paul.
Schaffer, Jonathan. 2003. Is there a fundamental level? *Noûs* 37(3): 498-517.
Schroeder, John. 2000. Nāgārjuna and the doctrine of "Skillful Means." *Philosophy East & West* 50(4): 559-583.
Sloman, Aaron. 2000. What are virtual machines? Are they real? http://www.cs.bham.ac.uk/~axs/ (accessed August 5, 2004).
Stcherbatsky, Theodore. 1962. *Buddhist Logic I & II.* New York: Dover.
—. 1923. *The Central Conception of Buddhism and the Meaning of the Word "Dharma."*
Varela, Francisco J. 1999. *Ethical Know-How: Action, Wisdom, and Cognition.* Stanford, CA: Stanford University.
Wallace, B. Alan. 1996. *Choosing Reality: A Buddhist View of Physics and the Mind.* Ithaca: Snow Lion.
—. ed. 2003. *Buddhism and Science: Breaking New Ground.* New York: Columbia University.
Wiener, Norbert. 1914. Relativism. *The Journal of Philosophy, Psychology and Scientific Methods* 11(21): 561-577.
—. 1961. *Cybernetics: or Control and Communication in the Animal and the Machine.* 2nd Ed. Cambridge: MIT Press.
Wood, Julia T. 2004. Buddhist influences on teaching and scholarship. *Journal of Communication and Religion* 27: 32-39.
A-kya Yong-dzin Yang-chän ga-wäi lo-dr'ö. 1976. *A Compendium of Ways of Knowing: A Clear Mirror of What Should be Accepted and Rejected.* Dharamsala: Library of Tibetan Works and Archives.
Zalher, ed. and Hopkins, trans. 1983. *Meditative States.* See Ch. 4, "Calm Abiding" from an oral presentation by Lati Rinbochay, edited by Leah Zahler and translated by Jeffrey Hopkins.

CHAPTER SEVEN

KNOWLEDGE AS A DUTY: THE ETHICAL SIGNIFICANCE OF THE INTEREST IN INFORMATION AND KNOWLEDGE

LORENZO MAGNANI
UNIVERSITY OF PAVIA

Knowledge as a Duty

Modern technology has brought about consequences of such magnitude that old policies and ethics can no longer contain them. For example, modern technology has made nature an object of human responsibility, and as a result, we must approach her not only with cleverness but also with ethics.

But as Hans Jonas has noticed (1974), as we seek to develop appropriate ethics, we can no longer expect to follow Immanuel Kant, who assumed that "there is not need of science or philosophy for knowing what man has to do in order to be honest and good, and indeed to be wise and virtuous" (Kant 1964, 72). Kant also says "[ordinary intelligence] can [...] have as good a hope of hitting the mark as any philosopher can promise himself" (Kant 1964, 72), and goes on to say that "I need no far-reaching ingenuity to find out what I have to do in order to possess a good will. Inexperienced in the course of world affairs, and incapable of being prepared for all the chances that happen in it," I can still ascertain how to act in accordance with moral law (Kant 1964, 70). Finally, Kant observes that even if wisdom "does require science," it "in itself consists more in doing and in not doing than in knowing," and so it needs science "not in order to learn from it, but in order to win acceptance and durability for its own prescriptions" (Kant 1964, 73). Jonas concludes: "Not every thinker in ethics, it is true, went so far in discounting the cognitive side of moral action" (Jonas 1974, 8).

Indeed this idea of knowledge as something simple and readily available to all people of good will is not sufficient anymore. In spite of Kant's declaration that wisdom and virtue are all one needs to make moral choices, decision making must be accompanied by *knowledge* that allows us to understand various problems and situations. And the knowledge involved should be not only scientific, but also human and social, embedded in reshaped "cultural" traditions. The "neighbour ethics" (Jonas 1974, 8) of justice, charity, honesty, etc., is no longer adequate for modern needs because such local action is overcome by collective action in which the effects are frequently not in a proximate sphere—consequences are often either spatially or temporally distant. To meet the new causal scale of individual and collective actions, a new long-range ethics of responsibility has to be built, one that is capable of dealing with the new global condition of human life.

As global stakes grow higher, knowledge becomes a more critically important *duty* than ever before and must be used to inform new ethics and new behaviours in both public policy and private conduct. This means that existing levels of intelligence and knowledge are no longer sufficient; today, ethical human behaviour requires that we assume long-range responsibility that is in keeping with the scope of our power to affect the planet, power that has grown exponentially over the years.

If we want knowledge to be considered a duty of the supernational society, then the goal should be to generate, distribute, and use knowledge to encourage economic and social development. Doing so, however, demands a great deal more research on creating and constructing knowledge for public and private use and, subsequently, on how to apply it to decision-making activities. The very future of the human species is at issue: if we accept the fact that it is our *duty* to construct both public and private bodies of knowledge, we will be more able to avoid compromising the conditions necessary for humanity to continue indefinitely on earth.

Unfortunately, at the beginning of the twenty-first century, we can see that in Western European and American societies, at least, knowledge and culture do not appear to be a priority. Even though there are funds, institutions, and human resources devoted to reproducing cultural activities, the vitality of culture remains in question. For example, we are confronted with a real breakdown in many humanistic traditions: in spite of its potential to help, research in humanistic fields—from philosophy to sociology to cognitive science—is voiceless before the problems of society. The intellectual exchange among these disciplines occurs mainly within universities and, as a result, creates a kind of academic short circuit. Intellectuals outside universities are similarly isolated, and their work is rarely woven into the fabric of society in general. It seems people who do not inhabit intellectual circles value research only if it relates to technology or to science related to technology, but such knowledge, unhappily, is not sufficient to build a knowledge-based society. Technology, while important, is blind to many problems and conflicts we will face in the future, and knowledge built only from scientific fields has limited value, as would knowledge drawn only from humanistic fields. Knowledge is a duty, but not just any knowledge: it must be a well-rounded, varied body of information drawn from many disciplines, and even then it must be useful, "available," and "appropriately applied."

Indeed, we must remember that even in rich countries, knowledge gleaned only from science and technology can be difficulty available and exploitable. For example, imagine a country with many atomic power plants, which require highly trained workers: the ability to generate adequate power, then, is directly related to the population's level of education and training. This relationship, in turn, affects global welfare and the organization of the country in question. A striking real-world example is the disaster at the Chernobyl nuclear power plant if seen in the light of the decline of Soviet Union and of the consequent weakening of agencies responsible for training workers at nuclear sites in particular and for heightening the public's awareness of technological and scientific issues in general. Of course in cases analogous to this, a lapse in scientific knowledge or training not necessarily is the only factor: ethical aspect—specifically, a lack of ethical knowledge—came into play as well. Moreover, even collectivities with highly qualified personnel and an advanced level of scientific-technological "know how" can be vulnerable to accidents if that knowledge is used "inappropriately" and not only if there is a failure of ethical (and, of course, political and legal) controls and knowledge.[37]

Rational Knowledge and Values

The rise of modern science has generated an increase in *rational* knowledge. In the last century, especially in Western Europe and the U.S., the fields of philosophy, logic, epistemology, and cognitive science have effectively mapped out the rational capacity of both human beings and their technological products. But a similar advance has not been made in ethics or in the ethical aspects of public policies and decision making; we have yet to deepen the exploration of the relationships between rationality and *values*. The "construction" of a modern ethical self-correcting knowledge is far from being achieved, yet it is critically important if we are to balance the needs of the poor and the rich around the world. In Europe, the historicist tradition has "culturally" almost killed the rationality derived from Galilean science and has generated various kinds of sterile (and dangerous) irrationalism. In the United States, overestimating the importance of productivity and technology is leading to a national collective

[37] My argumentation in this paper also takes advantage of Himma's considerations on the moral significance of the interest in information (Himma 2004).

consciousness that is largely devoid of cultural awareness: only technological products seem to have an effect on sub-culture and sub-cultural identity. For example many habits of everyday people have been shaped by corporate oriented mass-media advertising and stereotypes where the their message in terms of modern rationality is unfortunately jeopardized by their triviality; this fact has also favoured the growth of other kind of regressive identifications, for instance resorting to old-fashioned pseudo-religious convictions of medieval flavour.

We must use the current findings of philosophical, epistemological, cognitive, and human sciences—findings that support the idea of knowledge as a duty—to build new kinds of *rational traditions*. I think these traditions can be enhanced by reestablishing knowledge as a priority and then using it to guide social choices. In this way, we will be able to recover and restore some of the treasures of the Western world's cultural and scientific history.

It is worth noting that some transnational communities, those composed of "a network of professionals with recognized expertise and competence in a particular domain and an authoritative claim to policy relevant knowledge within the domain or issue area," are called *epistemic* communities (Hass 1990). Burkard Holzner and John Marx introduced the term and defined them as "those knowledge-oriented work communities in which cultural standard and social arrangements interpenetrate around a primary commitment to epistemic criteria in knowledge production and application" (Holzner and Marx 1979, 108). Holzner and Marx have received attention in the recent literature on international cooperation for their work that is also related to the management of the "commons." They describe epistemic communities that, for example, exist among Antarctic scientists, international marine lawyers, radio spectrum and telecommunications engineers (Vogler 2000, 28).

Transdisciplinary Issues

Creativity and model-based knowledge

Some transdisciplinary sub-topics mainly affect the central aspects of "knowledge-as-a-duty" in a multiple fashion. Thanks to the clarification provided by cognitive science and artificial intelligence, epistemological studies on creative and explanatory reasoning can bolster the argument for regarding knowledge as a duty. For example, abductive reasoning and model-based reasoning can be exploited to stress particular issues in scientific and ethical reasoning.

First of all we have to stress the importance of creativity: knowledge plays a creative role in attributing new values to external things. Building a body of rich and varied ethical knowledge that addresses today's problems is certainly a duty, just as it is to develop knowledge in other areas like science and medicine, for example. Enhancing our ethical wisdom requires work on several fronts: developing verbal/propositional and model-based knowledge; exploring manipulative and "through doing" knowledge; and learning to communicate effectively the results of epistemological studies and to incorporate the information into our everyday lives. And there are still more challenges created by the idea of "knowledge as a duty": dealing with its intrinsic "incompleteness" in various ethical and epistemological settings—ethical reasoning always occurs in presence of incomplete information; the effect of various inconsistencies (what can we do when we are faced with ethical issues that present contradictory aspects?); and the problem of comparing alternatives in theoretical and deliberative settings (see "Comparing alternatives and governing inconsistencies").

Knowledge and moral mediators in computational philosophy: distributing morality

The external objects and structures to which cognitive aspects and roles are delegated—blackboard with a diagram, for example—I call *epistemic mediators*. We can extend this idea beyond simple objects and structures, however, to include human organizations, which allows us to view these entities as *cognitive* mediators endowed with moral aspects. In a recent book on creative reasoning, I have described epistemic mediators not only as external objects and structures but also as human organizations—in this case, viewed as distributors of externalized cognitive potentialities (Magnani 2001a). Cognitive mediators function as enormous new external sources of information and knowledge, and, therefore, they offer ways of managing objects and information that cannot be immediately represented or found internally using only "minds." Analyzing these external structures is especially important in clarifying the role of media and of computational and information techniques; they are not merely mediators of information—they are mediators of "necessary" knowledge. Epistemic mediators also help to organize social and cognitive decisions made in academic settings, as is the case when. Examples of epistemic mediators are for instance artifacts in a scientific labouratory (a telescope or a magnetic resonance imaging machine) but also the collective of scientists itself, that is characterized by a specific distribution of cognitive roles, skills, and duties.

I think the best approach to studying these problems is to use so-called *computational philosophy*. The advent of certain machines and various rational methods and models brought about a computational turn in the last century, and this shift has revealed new ways to increase knowledge by embedding it in scientific and technological environments and by reshaping its major traditional topics (Magnani 1997). Just to make an example, the role of PCs and Internet in improving scientific research is very clear. In the new century, computational philosophy will allow an analysis of problems in recent logical, epistemological, and cognitive aspects of modelling activities employed in scientific and technological discovery. Computational philosophy supplies modern tools (new concepts, methods, computational programs and devices, logical models, etc.) to reframe many kinds of cultural (philosophical, ethical, artistic, etc.) knowledge that would remain inaccessible using old approaches, just mainly centered on the exploitation of mere “narratives”.

It is in this light that I introduce the concept of the *moral mediator*. Moral mediators play an important role in reshaping the ethical worth of human beings and collectives and, at the same time, facilitate a continuous reconfiguration of social orders geared toward rebuilding new moral perspectives. Finally, thinking in terms of human cognitive capacities, a person can be considered a kind of “thing” that can incorporate information, knowledge, know-how, cultural tradition, etc., just as cognitive objects like a book, a PC, or a work of art do. Unfortunately, human beings are sometimes assigned less value than things and moral mediators can help us redefine people as equally worthy of moral consideration.

Unexpressed knowledge, super-expressed knowledge, responsibility

As stressed by George Von Krogh, Kazuo Ichijo, and Ikujiro Nonaka, the need to elicit *unexpressed knowledge*—that is, isolated pockets of information in an organization that ought to be elicited more widely distributed—is a considerable challenge in knowledge management (KM), one that must be addressed as we take on the knowledge-as-a-duty problem (Von Krogh, Ichijo, and Nonaka 2000). KM helps generate assets for an organization by transferring people’s knowledge, expertise, and innovation, allowing them to be more effectively utilized; it leverages off an organization’s *intellectual capital* (people’s knowledge and expertise), and in combination with accurate, qualitative, and pertinent data, it provides an improved product. In KM, it is clear that what people know and can learn is more valuable than any other resource. Unexpressed knowledge becomes particularly important in the context of the opposite problem, what I call *super-expressed* knowledge: a huge and expensive cache of knowledge that is available but not used (and/or not useful). To make an example, a huge amount of explicit statistical data is constantly produced in the rich societies, theu will never be used or will just be used as a curiosity.

The problem of super-expressed knowledge is related to the vast collection of knowledge on the Internet and to the ethical question of privacy (cf. below). I maintain that while the concept of "knowledge as a duty" certainly applies to scientific, social, and political problems, it is an even greater factor in mapping ethical behaviours in the cyberage. The Internet provides a staggering amount of information that can influence human beings" welfare, and if we fully and universally acknowledge the Internet's immense power, we can immediately see that moral problems arise: issues of accuracy, deception, copyright, patents, trademarks abuses, and retention of information, just to name a few, have become much more complex as a result of the Internet. While these problems are vexing, I contend that "knowledge as a duty" will help guide us through this ethical labyrinth and reinforce the right to privacy (cf. below).

In a recent book I have also reconsidered the problem of human dignity in the era of globalization by analyzing mechanisms used to form rational ethical arguments (chapters six and seven) and by providing integrated analyses of current debates: cloning (chapter two), cyberprivacy (this chapter), hybrid people and consciousness (chapter three), and the environment (chapter one) (Magnani, 2005). Two common denominators of all these issues are an absence of appropriate and situated knowledge and, as a result, wide-spread cultural wariness and skepticism. Knowledge is abstract and general: cloning is treating people as means, privacy is needed just to preserve identity, preservation of nature is related to human beings' interests. The problem is that cloning can be in the same time treating people as means but necessary and useful in a particular case, preserving nature too dangerous to people in certain settings, the excess of privacy maintenance a threaten to the people's life. This skepticism clearly shows the most devastating effects in moral deliberations. Decisions are adopted following too general perspectives that do not take care of the concreteness of specific situations, or simply informed by extra-moral reasons given the unsatisfactory character of the available too abstract moral options. Furthermore, a lack of knowledge makes it almost impossible to behave responsibly in any situation.

As stated in earlier, human dignity is inextricably linked to the concepts of responsibility and freedom. I contend that both moral and scientific knowledge—when available, exploitable and appropriate—are powerful factors in developing responsibility and protecting freedom; the moral gap caused by bad faith results mainly from inadequate and "incomplete" assets of knowledge and information. This, in turn, generates such unbearable anxiety and anguish for human beings that they start to play a fictitious role. Enriching our knowledge about our "condition" is critically important if we are to fortify human freedom and to create incentives to adopt more responsibility.

New "knowledge communities:" humans as knowledge carriers

Technology is central to modern society, but those who create it are not taught to anticipate or investigate its impact, and, unfortunately, those who focus on technology's influence rarely understand the technologies themselves. I think that universities have a special responsibility, and therefore a unique opportunity, to address this issue. Especially since the nineteenth century, European and North American universities and technical institutions constructed a markedly bifurcated curriculum: one track is a technical/engineering/scientific curriculum, with courses in math, computing, and engineering taught in one intellectual region of the campus, while courses in the social sciences, arts, and humanities courses are taught in some other area. But the central questions in society are not about engineering or technology as such—rather, they encompass social, economic, political, and cultural issues. Taking on these questions will require the coordination of university curricula, a fruitful integration of engineering and liberal arts that might pair, say, bioprosthetics with disability literature. Also necessary is a partnership between research and service, and in this light the relationship between universities and industry is important. New *knowledge communities* have to be built. We must envision new alliances between universities, institutes, and business, and redefine our very idea of "knowledge" itself and the ways that knowledge is distributed. For example creative efforts are needed to shape new roles and interrelationships of knowledge in books, journals, databases, but also in social implementation of the same; in knowledge and information security, patent rights and licensing rights, etc.).

While technology provides many benefits, such as higher living standards, wider choice, more leisure time, and improved communications, it also brings with it both environmental and human risks. Technology alienates people from nature, concentrates political and industrial power in dangerous ways, and begets industrial nations that are capital-intensive rather than labour-intensive; the result is increased unemployment and an over-dependence on ostensibly neutral experts who in fact have interests and agendas of their own.

Though valuable in many ways, technology not always is benign, and it has the capacity to cause many kinds of damage. A list of negative effects related to technology can be given. It can lead to uniformity in a mass society, to narrow criteria for efficiency like fragmentation and specialization, manipulation and impersonal treatment). Another negative outcome is uncontrollability, where "separate" technologies from an interlocking system constitute a total, mutually reinforcing network that seems to lead a life of its own, like something self-perpetuating and all-pervasive). Finally the classical bad consequence in terms of alienation of the worker in Marxian sense is well-known (Barbour 1992).

Intellectuals have offered many antidotes to the ill effects of technology:

embracing a new ethics of responsibility; returning to the genuine engagement we experience when we focus on simple things rather than on production and consumption; recovering an imaginative and emotional life as well as religious commitment; redirecting technology according to social mores or religious constraints; and, finally, using new political projects to re-establish the importance of the personal and the individual as opposed to the materialistic and impersonal in society.

I think the role of human beings as knowledge carriers—as "biological" knowledge repositories, processors, disseminators, and users—is pivotal in our era of technology and globalization. In the globalised world, knowledge-carrying human beings enjoy increased status in various work collectives, social settings, and environments that have become more and more information-intensive, even if they are also increasingly disenfranchised and fragmented. Such people collabourate with (and, in some sense, are integrated with) non-humans objects and tools that manage and process a huge quantity of information. There are certainly many perils lurking in this globalized era—the alienation and exploitation of human beings come to mind, as does the obliteration of local cultures. But I think these new trends in technology also have the potential to benefit humankind in many ways: disseminating and universalizing information and eliciting unexpressed knowledge could bring exciting new possibilities and new opportunities for freedom and free choices.

Consequently, human beings, endowed by technology with new social intelligence, will be able to draw from their own reservoirs of knowledge and, at the same time, effectively wield new technology and artifacts in a way that heightens their intellectual and cooperative powers. As knowledge carriers, people are just as important as the non-human artifacts to which they are so closely linked; the human hybrid is, in certain respects, greater than the sum of its parts. Looking at human knowledge carriers in this way, it is easy to respect them as much as we respect, say, a computer, and this attitude will facilitate critical shift in perspective as we face the important task of learning to "respect people as things."

Identity and Privacy in the Cyberage

New moral ontologies

Cyber-warfare, cyber-terrorism, identity theft, attacks on abortion providers, holocaust revisionism, racist hate speech, organized crime, child pornography, and hacktivism—a blend of hacking and activism against companies through their web sites- are very common on the Internet (Van den Hoven 2000, 127). As a virtually uncontrollable medium that cuts across different sovereign countries, cyberspace has proved to be fertile ground for new moral problems and new opportunities for wrongdoing, mainly because it has created new moral ontologies that affect human behaviours.

More than in the past, many aspects of an information about human beings are now simulated, duplicated, and replaced in an external environment. Beyond supports of paper, telephone, and media, many human interactions are strongly mediated and potentially recorded through the Internet. At present, identity must be considered in a broad sense: the amount of data about us as individuals is enormous, and it is all stored in external things/means. This repository of electronic data for every individual human, at least those in rich countries, can be described as an external "data shadow" that, together with the biological body, forms a kind of cyborg that identifies us or potentially identifies us. The expression "data shadow" was coined in Sweden in the early 1970s: it refers to a true, impartial image of a person from a given perspective.[38]

When delegating information or tracing identity, human beings have always used external mediators such as pictures, documents, various types of publications, fingerprints, etc., but those tools were not simultaneously electronic, dynamic, and global, as is the case with the Internet. Where we exist cybernetically is no longer simple to define, and technologies of telepresence continue to complicate the issue.

New human beings: biologically local and cybernetically global

This complex new "information being" involves new moral ontologies. We can no longer apply old moral rules and old-fashioned arguments to beings that are both biological (concrete) and virtual, situated in a three-dimensional local space but potentially existing also as "globally omnipresent" information-packets (Schiller 2000). It is easy to materialize cybernetic aspects of people in three-dimensional local space, even if it is not quite as spectacular how Scottie beamed the glimmering forms of Captain Kirk and Mr. Spock from the transporter room to the surface of an alien planet in old *Star Trek* episodes. These aspects—being an agent, being an object, making an immoral or illegal act, etc.—are usually transparent when applied to macro-physical objects in local spatio-temporal coordinates. In cyberspace, however, they suddenly become obscure:

> What is an agent? What is an object? What is an *integrated product*? Is that one thing or two sold together? What is the *corpus delicti*? If someone sends an e-mail with sexual innuendo, where and when does the sexual harassment take

[38] Gotterbarn 2000: 216. Cf. also the amazing example of the "pizza profile" memorized in the pizza parlor caller ID, illustrated in Moor 1997.

> place? Where and when is it read by the envisaged addressee? Or when it was typed? Stored on the server? Or as it was piped through all the countries through which it was routed? And what if it was a forwarded message? (Van den Hoven 2000, 131).

New technologies threaten to increase our moral ignorance. I have previously pointed out, the idea of "knowledge as a duty" is relevant to scientific, social, and political problems, but it is even more important when we are faced with reinterpreting and updating the ethical intelligibility of behaviours and events. A problem of moral epistemology must be clearly addressed.

The new moral issues of the cyberage are well-known. Some problems, like financial deception and child abuse, were perpetrated for years via traditional media and have become exacerbated by the Internet. Others issues arose as a result of new technology—problems like hacking, cracking, and cooking up computer viruses, or the difficulties created by artificial intelligence artifacts, which are able to reason on their own (Steinhard 1999). Also needing attention is the proliferation of spam: its utility is negated when it is relegated to the recipient's email trash folder, and all it seems to accomplish is the externalization of cost for computer users.

Yet another challenge is the huge problem of the digital divide, for while millions of people are active cyborgs, a gulf has opened between them and others who number in the billions. Members of the latter group come from both rich and poor countries, and they have not attained cyborg status and/or are not aware they have a "data shadow." A further split, at least inside the "cyborg community," occurred because of the way technology dictates the speed, ease, and scope of communication; the dominance of certain text features creates an unlevel playing field, and the disparity between computer users with the most sophisticated equipment and those with more modest systems indicates that we are far from the ideal of an unmediated Internet. And the list continues: professional abuses, policy vacuums, intellectual property questions, espionage, a reduced role for humans—all these things create further moral tangles. Finally, if we consider the Internet as, technically speaking, a global collection of copying machines, like Mike Godwin contends, huge copyright problems arise (Godwin 1998, Johnson 1991, and Stallman, 1991).

In the light of the motto "knowledge as a duty," the Internet provides a lot of public critical goods for human welfare, by increasing access to information and knowledge, in terms of people's biological needs: food, drink, shelter, mobility, and sexuality. It also serves their rational aspects by providing information, stimulating imagination, challenging reason, explaining science, and supplying humor, to cite a few examples (Van den Hoven 2000). If the Internet's power, fecundity, speed, and efficiency in distributing information are universally acknowledged, moral problems immediately arise, related to distortion, deception, inaccuracy of data, copyright, patents, trademarks abuses.[39] It also seems that the Internet facilitates bureaucratic control: for instance in Africa it increases control over populations treated as the passive recipients of state policies.

Special affective relationships may lead to moral problems like identity theft as well as unfulfilled expectations or obligations. In both situations, freedom of speech must be protected—even if it sometimes means tolerating defamation, racism, hate speech, and violent pornography. Similarly, we must safeguard freedom of information and, at the same time preserve privacy, the right to private property, freedom of conscience, individual autonomy, and self determination. Technolibertarians say that VR (virtual reality) rapists (that is, men who simulate rape in virtual reality systems) are "of course assholes, but the presence of assholes on the system is a technical inevitability, like the noise in a phone line."[40]

Even in the complicated world of cyberspace, however, Mill's principle still establishes the moral threshold: only the prevention of harm to others justifies the limitation of freedom. He also observes that the truth emerges more easily in an environment in which ideas are freely and forcefully discussed. It is well-known that the Communications Decency Act passed by the United States Congress in 1996 had a very short life and was rapidly struck down by the US Supreme Court.[41]

Cyberprivacy

We have said that the post-human "cyborg" comprises a vast quantity of external information and that it is possible to "materialize" at least many of our cybernetic aspects (ideally I can print at home your picture, your sexual, political, and literary preferences, what you bought in the last two weeks etc.). Consequently, those aspects are potentially "globally omnipresent" and not only available to us: everyone can, in theory, see them. We must note that computer technology also distances the human body from the actions performed, and because it is an automating technology, it has a great potential to displace human presence.

---

[39] Goldman 1992; Mawhood and Tysver; 2000. On the Internet's retention of information in corporations, governments, and other institutions, cf. Davis and Splichal 2000.

[40] Dibbell 1993. Some recent perspectives on the ethics of the "meat-free existence" in cyberfeminism and on the problem of "women and technology" are given in Flanagan and Booth 2002, and in Adam 2002.

[41] On the destiny of free speech in the digital age, cf. Godwin 1998. Many data on the problem of the digital divide in the US and Europe are given in Compaine 2001.

The possibility of materialize the aspects of human being is related to the so-called Panopticon effect. The Panopticon, designed by Jeremy Bentham, was an architectural design for prisons that allowed just a few guards to keep an eye on a great many convicts. The structure was a round, multi-story building with each floor's cells arranged in a circle around a hollow core. Rising from ground level in the center, in a sort of round atrium, was a guard tower with windows facing out toward the cells. Because each cell was illuminated by a window to the outside, and the cell's inner wall consisted only of bars, guards could easily monitor the prisoners, every one of whom was completely visible at all times. Michel Foucault described the Panopticon as a metaphor of the "mechanisms of large-scale social control that characterize the modern world" (Foucault 1979; Reiman 1995). Of course, a person's being on display is enough by itself to effect social control: many psychological studies have observed that when people believe they are visible to others, they behave differently from the way they do when they believe they are out of view. Foucault considered the practices of medicine, psychology, and sex education to be forms of social control "because they create a world in which the details of our lives become symptoms exposed to a clinical *gaze*—even if not actually looking" (Reiman 1995, 29)

The Internet can be seen as a modern Panopticon that makes us all visible to anyone in a position to see. There is a tremendous amount of information on the net: on display are census data, addresses, credit card purchases, credit histories, medical information, employment histories, evidence of intimate relationships, indicators of religious inclinations—the list could go on and on. But unlike Bentham's prison, where prisoners could be watched only from a single point—the central guard tower—we can be "viewed" from billions of sites, for all it takes is a computer with an Internet connection. Computer networks have no centers, and almost any portion can operate as an autonomous whole, which is also what makes controlling the network so difficult. As I have already illustrated, a detailed computational "shadow" of a private life can easily be built by collecting information that is available online.

The availability of too much personal information increases our vulnerability to all kinds of dangers, so protective measures must be instituted; on the other hand, however, monitoring the communications of others contradicts the universal right to privacy. Although there are some cases when too much privacy can be catastrophic—when families hide abuse against women and children, for instance—the cyberage renders privacy increasingly important. Heightened privacy poses its own difficulties, of course, bringing challenges that include greater costs as well many new needs—the need for developing new ethical and technical knowledge, establishing new political priorities, and passing new laws. There are those who maintain that the more privacy we have, the harder it is for society to gather the information needed to apprehend and punish criminals, although I am skeptical about that assertion. Members of society must be protected not only from being seen but also from feeling visible, and both conditions allow us to see privacy either as a basic right that some people are deprived of, as in an inaccessibility to information about ourselves or, even more important, as a method of controlling who may have access to us (Fried 1984). Some ethicists contend this second idea of privacy leads to anomalies: for example, the moral right to privacy has to be limited to "my" right that others be deprived of the access to me. Nevertheless, we certainly want to establish laws that protect privacy so that it is guaranteed to be not only a moral right, but a legal one as well, even (and especially) in the era of the so-called "Information Panopticon" (Fried 1984, 32-34; Edgar 1997; and Tribe 2000).

The Internet and the so-called "Intelligent Vehicle Highway Systems" (IVHS) have created new external devices that challenge our established rights to privacy, making it necessary to clearly define and shape new moral ontologies. Recently, ECHELON, the most powerful intelligence-gathering organization in the world, has garnered attention as it has become suspected of conducting global electronic surveillance—mainly of traffic to and from North America—which could threaten the privacy of people all over the world. The European Union parliament recently established a temporary committee on the ECHELON interception system (1999-2004) to investigate its actions.[42]

Because so much information has become available about people, the aim must be to anticipate and avoid possible negative consequences. The ostracism and stigmatization of minorities could increase and generate an explicit loss of freedom. Moreover, less choice may exist because of an implicit loss of freedom: to make an example "I am no longer free to simply drive to X location *without leaving a record,*" an idea explored further below. Less privacy could also mean the loss of protection against insults that "slight the individual's ownership of himself" (as happens in the case of prisoners or of slaves): as the consequence people will say *mine* with less authority and *yours* with less respect. Indeed when we lose ownership of our actions, we lose responsibility for them, given the fact that morality—as I contend—involves the right to own our own destinies and to choose in our own way. Finally, the risk increases of infantilizing people and making them conformist and conventional, and consequently, the possibility increases of their being oppressed. At this point it is patent that, given the items above, privacy is politically linked to the concepts of liberalism and democracy (Reiman 1995, 35-44).

[42] The 2001 draft report "on the existence of a global system for the interception of private and commercial communications (ECHELON interception system)" can be found at http://www.europarl.eu.int/tempcom/echelon/pdf/prechelon_en.pdf.

Philosophy wonderfully stated the fundamental character of privacy with the clear word of John Stuart Mill: "The only part of the conduct of anyone, for which he is amenable to society, is that which concerns others. In the part which merely concerns himself, his independence is, of right, absolute. Over himself, over his own body and mind, the individual is sovereign."[43]

Most authors agree that a cyberdemocracy must be established to emphasize the role of civil rights and responsibility in cyberspace. Johnson's argumentation about the role of Internet technology in promoting democracy concludes in a enigmatic way: the Internet "has the capacity for hegemonic as well as democratic relationships." He goes on to pose many provocative questions: 1) how is Internet development related to the digital divide? 2) Does the Internet move towards private interests and market focus? 3) Can the participation it guarantees influence decision-making processes? 4) Does the Internet threaten the autonomy and democratic status of nations? 5) To what extent is there privacy on the Internet, and to what extent is it a medium of surveillance? It seems Internet is neither in the service of enhancing democracy nor of promoting global democracy.[44]

Finally, we can distinguish two kinds of privacy violations: the first involves disseminating intimate information to a general audience without the consent of the person involved,[45] and the second occurs when that personal information is used to make decisions concerning the individual without his involvement, possibly to harm him. There is always a conflict between privacy and freedom of information. Medical information, while rarely disseminated, can become emotionally charged, for a person may feel lost, unprotected, and vulnerable if the information is in fact spread. An example in which health records may be communicated to others is the case of epidemiological research, which is possible only when medical information about subjects is gathered and examined by scientists, but even in this case, each participant's privacy can be

---

[43] Mill 1966: 14. Other aspects and benefits of privacy like promoting mental health and autonomy are illustrated in Gavison 1984.

[44] Johnson 2000: 190-196. Recently information ethics scholars have increased knowledge about many of these and other topics: cf. for example the classical Maner 1980; Moor 1985; Gotterbarn 1991; Rogerson 1996; and Johnson 1994; more recent, Floridi 1999; 2002; 2004; Bynum 2004; 2005 (on Norbert Wiener's vision of the information society), Floridi and Sanders 2003 and 2004, Moor and Bynum 2002; Bynum and Rogerson 2004.

[45] Elgesem 1996. On the role of cookies, sniffers, firewalls, data encryption, filtering systems, digital signatures, authentication, and other technical tools to threaten or to protect privacy, cf. Beckett 2000; Weckert 2000; and Spinello 2000. EU countries introduced Data Acts to protect data concerning people in 1973 and 1974. The US still lacks this legislation, its government prefers to promote self-monitoring and self-regulation.

protected by employing pseudonymisation. It really does seem that privacy can be considered both an intrinsic and an instrumental value essential to democracy, autonomy, and liberty (Tavani 2000, Johnson 1994).

Privacy, intimacy, self-deception, and the ownership of our destinies

The lack of privacy surely injures one's self esteem and assaults his feeling of "human dignity" as long as this lack is a clear threat to his individual "liberty" (Bloustein 1984, 160). We have said that the post-human "cyborgs" are designated as such because they are composed of a huge quantity of electronically incorporated external information and identity. Externalizing information about ourselves in "things" that are just "over there," which, of course, makes that information available to others, makes possible the loss of very important ends like respect, love, friendship, and trust. These concepts, through which we express some of the more sophisticated aspects of human animals, are simply inconceivable without privacy (Fried 1984, 205).

If we consider privacy to be the ability to control information and knowledge about ourselves, it is obvious—as illustrated by the Panopticon effect—that we are less free when we are observed or if we believe we are being observed. Electronic monitoring drastically reduces the power we have over information about ourselves: "monitoring obviously presents vast opportunities for malice and misunderstanding on the part of the authorized personnel. For that reason the subject has reason to be constantly apprehensive and inhibited in what he does" (Fried 1984, 216).

Moreover, because intimacy is the conscious sharing of information about ourselves that is typically not shared at all, privacy is the means to create what has been called the "moral capital" that we spend, for example, in love and friendship. Privacy grants control over information so that we can preserve intimacy. Even between husband and wife, the restraints of privacy hold; although marriage implies a voluntary renunciation of private information, one spouse never knows what the other has chosen not to share.

If we consider trust to be the expectation that another person will behave morally, privacy becomes essential: there cannot be trust where there is not a possibility for error. To foster trust, not all information need be shared: a person cannot know that she is trusted unless she has the right to act without constant surveillance so that she must be free to choose to honor or betray the trust, and it is privacy that furnishes that fundamental right.

In summary, in order to engage in intimate relationships like love, friendship, and trust, we must reveal some of ourselves, but if this information is seen by too many others, it loses its intimate nature so that love, friendship, and trust are threatened. It now appears clear that if we are deprived of the control over information about ourselves this reflects—because of the effect on love, friendship, and trust—an assault to our liberty, personality and self-respect.

Something more must be said about the deleterious effects of the loss of privacy. The right to privacy is also related to the idea of respecting other people as "choosers": "To *respect* someone as a person is to concede that one ought to take account of the way in which his enterprise might be affected by one's own decisions. By the principle of respect for persons, then, I mean the principle that every human being, insofar as he is a qualified person, is entitled to this minimal degree of consideration" (Benn 1984, 229).

Consequently, privacy can be defined as the protection against interference with a person's way of realizing and developing her interests, both in terms of practical objectives or of merely shaping one's own identity over time. By protecting privacy, we protect the ability to develop and realize projects in our own ways. If respect for people is related to the respect for their rights, privacy becomes paramount. A monitored individual is indeed inclined to "give himself away" (Fried 1984, 218), and so he is compelled to appear false, cold, and unnatural. This can affect the attitude of other people whose esteem, love, and friendship he desires. This kind of humiliating auto-betrayal has something in common with the bad-faith condition I have describe in a recent book (Magnani 2005). Both conditions share an externalization of responsibility and freedom, and thus a decrease of personal human dignity.[46]

References

Adam, A. 2002. The ethical dimension of cyberfeminism. In *Reload. Rethinking Women + Cyberculture*, eds. M. Flanagan and A. Booth, 159-174. Cambridge, MA: MIT Press.

Barbour, I. 2001. Views of technology. In *Computers and Ethics in the Cyberage,* eds. D. Micah Hester and P. J. Ford, 7-34. Upper Saddle River, NJ: Prentice Hall.

Beckett, D. 2000. Internet technology. In *Internet Ethics,* ed. D. Langford, 13-46. New York: St. Martin Press.

Benn, S. 1984. Privacy, freedom, and respect for persons. In *Philosophical Dimensions of Privacy: an Anthology,* ed. F. Shoeman, 223-244. Cambridge: Cambridge University Press.

[46] Moor (1997) adds another justification of privacy, beyond the one in terms of control/restricted account I have adopted here: privacy is also justified as an expression of one of the core values – values that all normal humans and culture need to survive: life, happiness, freedom, knowledge, ability, resources, and security, In this sense privacy is justified as an essential member of the central framework of values for a computerized society.

Blounstein, E. J. 1984. Privacy as an aspect of human dignity, In *Philosophical Dimensions of Privacy: an Anthology,* ed. F. Shoeman, 158-202. Cambridge: Cambridge University Press.

Bynum, T. W. 2004. Ethical challenges to citizens of 'The Automatic Age:' Norbert Wiener on the Information Society. *Journal of Information, Communication and Ethics in Society* 2:65-74.

—. 2005. Norbert Wiener's vision: the impact of the 'automatic age' on our moral lives. In *The Impact of the Internet on Our Moral Lives*, ed. R. Cavalier, 11-25. Albany, NY: State University of New York Press.

Bynum, T. W. and S. Rogerson, eds. 2004. *Computer Ethics and Professional Responsibility.* Malden, MA: Blackwell.

Compaine, B. M. 2001. *The Digital Divide. Facing a Crisis or Creating a Myth?.* Cambridge, MA: MIT Press.

Davis, N. D. and S. L. Splichal. 2000. *Access Denied. Freedom of Information in the Information Age.* Ames, IA: Iowa State University Press.

Dibbel, J. 1993. A rape in the cyberspace. In *Computers and Ethics in the Cyberage,* eds. D. Micah Hester and P. J. Ford, 439-452. Upper Saddle River, NJ: Prentice Hall.

Edgar, S. L. 2000. *Morality and Machines: Perspectives in Computer Ethics.* Sudbury, MA: Jones and Bartlett.

Elgesem, D. 1994. Privacy, respect for persons, and risk. In *Computers and Ethics in the Cyberage,* eds. D. Micah Hester and P. J. Ford, 256-277. Upper Saddle River, NJ: Prentice Hall.

Flanagan, M. and A. Booth, eds. 2002. *Reload. Rethinking Women + Cyberculture*, Cambridge, MA: MIT Press.

Floridi, L. 1999. *Philosophy and Computing.* London: Routledge.

—. 2002. Information ethics: an environmental approach to the digital divide. *Philosophy in the Contemporary World* 9(1):39-45.

—., ed. 2004. *Blackwell Guide to Philosophy of Computing and Information.* Malden, MA: Blackwell.

Floridi L. and J. W. Sanders. 2003. The method of abstraction. In *Yearbook of the Artificial. Nature, Culture, and Technology. Models in Contemporary Sciences*, ed. M. Negrotti. Bern: Peter Lang.

Floridi L. and J. W. Sanders. 2004. On the morality of artificial agents. *Minds and Machines* 14: 349-379.

Foucault, M. 1979. *Discipline and Punish: the Birth of the Prison* (1975), trans. A. Sheridan. New York: Vintage Books.

Fried, C. 1984. Privacy. In *Philosophical Dimensions of Privacy: an Anthology,* ed. F. Shoeman, 202-222. Cambridge Cambridge University Press.

Gavison, R. 1984. Privacy and the limits of law. In *Philosophical Dimensions of Privacy: an Anthology,* ed. F. Shoeman, 346-402. Cambridge: Cambridge

University Press.
Godwin, M. 1998. *Cyber Rights. Defending Free Speech in the Digital Age.* Toronto: Random House.
Goldman, A. 1992. *Liaisons. Philosophy Meets the Cognitive and Social Sciences.* Cambridge, MA: MIT Press.
Gotterbarn, D. 1991. Computer ethics: responsibility regained. *National Forum: The Phi Beta Kappa Journal* 71: 26-31.
—. 2000. Virtual information and the software engineering code of ethics. In *Internet Ethics*, ed. D. Langford, 200-219. New York: St. Martin Press.
Haas, E.B. 1990. *When Knowledge is Power: Three Models of Change in International Organization.* Berkeley, CA: University of California Press.
Himma, K.E. 2004. The moral significance of the interest in information: reflections on a fundamental right to information, American Conference Computing and Philosophy, Pittsburgh, August 4-6, 2004, manuscript.
Holzner, B. and J. H. Marx. 1979. *Knowledge Application: the Knowledge System in Society.* Boston: Allyn and Bacon.
Johnson, D. G. 2004. Proprietary rights in computer software: individual and policy issues. In *Computer Ethics and Professional Responsibility,* eds. T. W. Bynum, and S. Rogerson, 285-293. Malden, MA: Blackwell.
—. 1994. *Computer Ethics.* 2nd Ed. Englewood Cliffs, NJ: Prentice Hall.
—. 2000. Democratic values and the Internet. In *Internet Ethics*, ed. D. Langford, 181-199. New York: St. Martin Press.
Jonas, H. 1974. Technology and responsibility: reflections on the new tasks of ethics. In *Philosophical Essays: from Ancient Creed to Technological Man,* 3-30. Englewood Cliffs, NJ: Prentice-Hall.
Kant, I. 1964. *Groundwork of the Metaphysics of Morals* (1785), ed. and trans. H. J. Paton. New York: Harper & Row.
Langford, D., ed. 2000. *Internet Ethics.* New York: St. Martin Press.
Magnani, L. 1997. *Ingegnerie della conoscenza. Introduzione alla filosofia computazionale.* Milan: Marcos y Marcos.
—. 2001a. *Abduction, Reason, and Science. Processes of Discovery and Explanation.* New York: Kluwer Academic/Plenum Publishers.
—. 2005. *Ethics and Cognition. Distributed Morality in a Technological World,* forthcoming.
Maner, W. 1980. *Starter Kit in Computer Ethics.* Helvetia Press (published in cooperation with the National Information and Resource Center for Teaching Philosophy).
Mawhood, J. and D. Tysver. 2000. Law and internet. In *Internet Ethics*, ed. D. Langford, 96-126. New York: St. Martin Press.
Micah Hester, D. and P. J. Ford, eds. 2001. *Computers and Ethics in the Cyberage.* Upper Saddle River, NJ: Prentice Hall.

Mill, J. S. 1966. *On Liberty* (1859). In *On Liberty, Representative Government, The Subjection of Women,* 5-141. Oxford University Press.

Moor, J. H. 1985. What is computer ethics? *Metaphilosophy* 16(4): 266-275.

—. 1997. Towards a theory of privacy in the information age. *Computers and Society* 27: 27-32.

Moor, T. H. and T. W. Bynum, eds., 2002. *Cyberphilosophy.* Malden, MA: Blackwell.

Reiman, J. H. 1984. Privacy, intimacy, personhood. In *Philosophical Dimensions of Privacy: an Anthology,* ed. F. Shoeman, 300-316. Cambridge Cambridge University Press.

—. 1995. Driving to the Panopticon: a philosophical exploration of the risks to privacy posed by the highway technology of the future. *Computer and High Technology Law Journal* 11(1): 27-44.

Rogerson, S. 1996. The ethics of computing: the first and second generations. *The UK Business Ethics Network News.* Spring: 1-4.

Schiller. 2000. The global information highway. In *Society, Ethics, and Technology*, eds. M. E. Winston and R. D. Edelbach, 171-181. Belmont: Wadsworth/Thomson Learning.

Shoeman, F., ed. 1984. *Philosophical Dimensions of Privacy: an Anthology*, Cambridge University Press, Cambridge.

Spinello. R. A. 2000. Information integrity. In *Internet Ethics*, ed. D. Langford, 158-180. New York: St. Martin Press.

Stallman, R. 2004. Why software should be free. In *Computer Ethics and Professional Responsibility,* eds. T. W. Bynum and S. Rogerson, 294-310. Malden, MA: Blackwell.

Steinhart, E. 1999. Emergent values for automations: ethical problems of life in the generalized internet. *Journal of Ethics and Information Technology* 1(2): 155-160.

Tavani, H. 2000. Privacy and security. In *Internet Ethics*, ed. D. Langford, 65-95. New York: St. Martin Press.

Tribe, L. H. 2000. The constitution in cyberspace. In *Society, Ethics, and Technology*, eds. M. E. Winston and R.D. Edelbach, 223-231. Wadsworth/Thomson Learning, Belmont.

Van den Hoven, J. 2000. The Internet and varieties of moral wrongdoing. In *Internet Ethics*, ed. D. Langford, 127-157. New York: St. Martin Press.

Vogler, J. 2000. *The Global Commons. Environmental and Technological Governance.* Chicester: Wiley.

Von Krogh, G., K. Ichijo, K., I. Nonaka. 2000. *Enabling Knowledge Creation. How to Unlock the Mistery of Tacit Knowledge and Release the Power of Innovation*, Oxford University Press.

Weckert, J. 2000. What is new or unique about Internet activities? In *Internet*

Ethics, ed. D. Langford, 47-64. New York: St. Martin Press.
Winston, M. E. and R. D. Edelbach, R.D. 2000. *Society, Ethics, and Technology*. Belmont: Wadsworth/Thomson Learning.

CHAPTER EIGHT

A COMPUTER ENHANCED GOOD LIFE

JOHN WECKERT
CENTRE OF APPLIED PHILOSOPHY AND PUBLIC ETHICS,
CHARLES STURT UNIVERSITY

The Good Life

We all have some idea of what a good life is for us, although we may all have slightly different conceptions of it. For Aristotle the one thing that everyone wanted for its own sake was happiness and the best way of achieving this was through a life of contemplation, or study. Happiness (or *eudaimonia*) for Aristotle was a much richer concept that it currently is for us. It involved living well and doing well, it was something that lasted for a sustained period (one could not be happy for just an hour), and it could not apply to just one aspect of life. In general it meant overall success, prosperity, and achievement. It resulted from virtue, but also required external goods, for example good birth, good children and beauty.

The happy person was the person who enjoyed a good life and as mentioned, this was only available to a person of virtue, who was one who avoided extremes and always chose the middle way between extreme positions or actions. Friendship was an essential component of this life (and true friendship is only possible for the virtuous person).

This summary of Aristotle's account of the good life is vastly oversimplified, but the important point here is that it is based on the kinds of beings we are, or at least the kinds of beings that he thought we are. The kind of life that he advocated was the only one that could lead to true happiness for us as we are but not necessarily for different creatures. His emphasis was on what constitutes a good life for a *person*, and ethics is the science of trying to discover this; a study of what people are like and what gives them happiness.

James Moor argues that there are certain core values that all people have (Moor 1999, 66). These are: life, happiness (pleasure) and autonomy. In order to exercise our autonomy we require the *ability* to do various things, the *security* to do them, *knowledge* about doing them, the *freedom* and *opportunity* to do them, and finally the *resources* to accomplish our goals. In order to achieve a good life we require all of these, but different people will not give all of the components the same weightings and therefore conceptions of the good life will be different for different people, but not radically so. Some people place more weight on security while others on freedom, for example, but all want some of each.

Again what is important is not so much the details of Moor's account but rather the fact that it is an account based on humans as we are. If we were different in relevant ways our path to a good life would be different. We might not be interested in freedom for example, or in security.

Technology and the Good Life

When we think of technology we tend to think of things like computers, cars and mobile telephones. Spears and stone axes are of course just as much part of technology and so, arguably is the use of fire. According to Ortega Y Gasset, technology can be defined as "… the improvement brought about on nature by man for the satisfaction of his necessities", and "Technology is the reform of nature" (1961, 95). Perhaps it is more common for us to think of technology as the tools and techniques for bringing about these improvements, but that does not alter Ortega Y Gasset's point.

He goes on to argue that "Not being, but well-being, is the fundamental necessity of man…", and therefore a main aim of technology is "to promote good life, well-being, by adapting the medium to the will of the individual" (1961, 101).

This too is an oversimplification of Ortega Y Gasset's position, but for our purposes his important point is that we humans use technology to modify our environment in order for us to live better lives, and perhaps even the good life.

Computer Technology and the Good Life

Just as technology in general is supposed to enhance our lives, so is computer technology. Initially stand alone computers helped us in perform mathematical calculations, store and retrieve data, write articles and so on. Later with the development of widespread computer networks, and especially the internet, they helped us communicate with one another just about anywhere in the world. And now computer processors are embedded in a vast variety of products, including cars, washing machines and telephones.

While there is room for the view that much of this development was and still is driven by the technology rather than by any human needs, and some cynicism is not out of place, it is plausible to see these developments as efforts to make life easier and more pleasant, keep us healthier and more generally to help us satisfy our goals and to enhance our lives. The extent to which our lives have been enhanced by computer technology may be debatable, but there is no doubt that our lives have been changed by it and that in many ways our lives are easier, so to that extent at least they have been enhanced. And they have been enhanced by changes to our environment.

A Break With the Past?

Computers, as we have just seen, have already and are continuing to change our environment, making it, arguably more pleasant for us. But now there is the possibility that computers will play a part in changing us. In a sense this is not new. We are accustomed to prosthetic limbs, spectacles, dentures and more recently, cochlear implants. These all modify the person is some way, but in a way that makes him or her more or less the same as other people with respect to some particular deficiency, for example poor sight or hearing. While individuals were changed there was no change in humans as a whole. We still had the same abilities.

But this could change. There is currently much discussion on Converging Technologies, which are commonly said to comprise Nanotechnology, Information Technology, Biotechnology and Cognitive Science. (Roco and Bainbridge 2002) These technologies taken together have the potential to quite radically enhance the abilities of humans. In the future, technology may not so much change our environment as change us. We will consider a few relevant developments that may be primitive relative to future developments, but they do demonstrate the direction things could take.

First is the remote controlled rat, or *ratbot*. This rat was controlled by inputs directly to its brain and to its whiskers. When its whiskers were stimulated and it performed the appropriate action the reward-centres of its brain were stimulated "filling the rodents with a feeling of well-being". So in effect, the scientists could control the behaviour of the rat by controlling its wants. That is, the rat was doing what it wanted to do, but the scientists were in control of what it wanted (Whitehouse) The second example is that of Kevin Warwick who was able to have direct communication through implants in his arms to the outside world, including with his wife who also had implants. Warwick is optimistic about the future of cyborgs, creatures that are part human, part machine, and who will have "… extra-sensory capabilities, a high-performance means of communication and the best of human and machine brains" (Warwick, 4). Both of these cases can be seen as early steps to new forms of human sensing, communication and reasoning ability.

Other research is also proceeding, particular at DARPA, the Defense

Advanced Research Projects Agency, of the USA, where there are a number of projects designed to enhance the abilities of people in the military (Goldblatt 2002). One of these aims to augment cognition, that is, it aims to give people an enhanced cognitive ability. Another involves research into the brain-machine interface. This involves research into giving people the ability to directly control devices by the brain, without having to use hands or feet or any other part of the body. For example, a pilot may be able to direct fly an aeroplane by the use of his or her brain, without physically needing to handle the controls. Other research is examining the possibility of the direct input of skills and information without the need for learning in the way that we learn now. Our brain perhaps could receive its download while we slept. In the morning we would know a lot that we did not when we went to sleep.

Human Enhancements and the Good Life

These types of developments raise two questions. First, are these enhancements likely to promote the good life? And second, will we need to rethink the whole notion of a good life in the light of them?

First, are enhancements likely to make life better for us? An enhancement here is taken to mean some change or modification made to us that in some way gives us abilities beyond what a human normally has. It might be vision as good as an eagle or hearing as good as a dog. Or it could be the ability to directly communicate with machines or with other people through transmitters and receivers connected to our brains. We are not talking here about treatments, modifications or devices that bring people up to what is more or less normal in the total population.

It is difficult to know whether enhancement would make life better overall. Consider direct communication between brains. How would this affect our privacy? Currently my thoughts are mine alone, more or less, something for which I am very grateful. This is not so much because I intentionally think things that I should not, but more that I do not have the same sort of control over my thoughts that I have over my actions. My autonomy would be reduced to the extent that I would have much less control over what people knew about me, and this would certainly make me more vulnerable to government and employer control.

Consider too enhanced senses. Suppose that I had the vision of an eagle. This may improve my life, I really cannot say because I have no idea what it would be like. But this raises another issue. If some of us have enhanced vision but not everyone does, how will this affect communication and general social cohesion? Perhaps this is not a problem. Those with visual and hearing impairments lack senses that most have while some of their other senses are more highly developed. But we can still communicate, even if this communication is limited in some respects, for example with respect to colour in the case of the visually impaired. The situation does not seem so different in the case of someone who may be enhanced by the addition of a different sense, say echolocation, like a bat. Presumably there would still be much about which we could communicate easily. But suppose that it became possible to have enhancements of many kinds and that we could choose which we wanted, if any. Could I, as a conservative, unenhanced human who is fairly happy being as he is, communicate easily with my highly enhanced neighbours who had all kinds of experiences of which I could not even conceive? Probably not. So while bodily enhancements may improve life in some ways, in others there seems to be a high price to pay if communication becomes difficult. Humans are basically social creatures who like living in groups, and this aspect would appear to be under threat. Could I be a friend of an enhanced person? Aristotle would probably say not. Real friends must be equals. Could I understand this person? Wittgenstein would have his doubts. After all, he did say that if a lion could talk we could not understand it (Wittgenstein 1958, 223). While both Aristotle and Wittgenstein might both appear to overstate the case, communication and friendship are easier between people who are similar.

None of this may matter of course. We, as we currently are, are social creatures, but our enhanced heirs may not be, or not in the way that we are. They may not mind if there are few with whom they can easily communicate. This takes us back to Aristotle and Moor. For both of them the conditions necessary for a good life are predicated on what people are actually like. If enhanced people are no longer like us, then of course we need to look afresh at what a good life might be.

Up until now we have had a pretty good idea of what makes life better and we can be fairly confident about what makes life better for most. In the future, with human enhancements, it will not be so clear. Do we know if particular "enhancements" will improve life? Will enhanced people be happier, and if not, why bother with enhancements? Can we say much about the "good life" for an "enhanced" person? One important factor is this. Currently people around the world are more or less the same. We know in general what sorts of things make people happy, what makes them suffer, what gives pleasure and pain, and so on. If human enhancements become widespread it is likely that people will become very different from each other. Many different kinds of enhancements, and enhancements to different levels may be possible, and if people are free to choose, they will choose differently. So it may not be known what should be done to relieve the suffering or increase the pleasure of others. In an important respect our commonality will be lost, and if this is so, the notion of "the good life" becomes vacuous in the sense that it can play no role as a goal for technological development. It was noted earlier that for Ortega Y Gassett the goal of technology was "to promote good life, well-being, by adapting the medium to the will of the individual". In a world of human enhancement this no longer holds. Technology no longer adapts just the environment but also the human, so our frame of reference, namely a common humanity, is gone. This may not matter, but it should make us reconsider the purpose of technological development. What will be a computer enhanced good life?

References

Aristotle. 1999. *Nicomachean Ethics*, trans. Terence Irwin. 2nd ed. Indianapolis, IN: Hackett Publishing Company.

Goldblatt, Michael. 2002. DARPA's programs in enhancing human performance. In *Converging Technologies for Improving Human Performance, Nanotechnology, Biotechnology, InformationTechnology and Cognitive Science,* eds. Mihail C. Roco and William Sims Bainbridge, 337-341. NSF/DOC sponsored report.

Moor, James. 1999. Just consequentialism and computing. *Ethics and Information Technology*, 1: 65-69.

Ortega Y Gasset, Jose. 1961. Man the technician. In *History as a System: And Other Essays Toward a Philosophy of History*, trans. Helene Weyl, 87-161. New York: W. W. Norton & Company.

Roco, Mihail C., and Bainbridge, William Sims. 2002. *Converging Technologies for Improving Human Performance, Nanotechnology, Biotechnology, InformationTechnology and Cognitive Science,* NSF/DOC sponsored report, http://wtec.org/ConvergingTechnologies/.

Warwick, Kevin. 2002. *I, Cyborg*. Century.

Whitehouse, David. Here come the ratbots (1 May 2002), http://news.bbc.co.uk/1/hi/sci/tech/1961798.stm.

Wittgenstein, Ludwig 1958. *Philosophical Investigations*, trans. G. E. M. Anscombe. Oxford: Blackwell.

CHAPTER NINE

KANTIAN ROBOTICS: BUILDING A ROBOT TO UNDERSTAND KANT'S TRANSCENDENTAL TURN

LAWRENCE M. HINMAN
DEPARTMENT OF PHILOSOPHY, UNIVERSITY OF SAN DIEGO

Introduction

Teaching Kant's first *Critique* can be notoriously difficult. The Kantian version of the Copernican revolution challenges students to think in new ways about human subjectivity and the nature of objectivity, and Kant's language certainly challenges them to read with an extraordinarily high level of attention to detail. Indeed, for many, the language can be so daunting that they simply give up before they get to the good part—the intrinsic rewards of mastering a difficult philosophical text.

Here is a *Gendankenexperiment*, a thought-experiment, which may prove helpful in drawing students into the basic arguments of the Transcendental Analytic in the *Critique of Pure Reason*. Thinking through the process of building a robot forces us to have many of the insights that Kant develops in the Transcendental Analytic and the first part of the Transcendental Dialectic. The purpose of this analogy is to help students understand the transcendental turn, not to advance Kant scholarship, although I think the analogy raises some interesting questions about Kant's position.

Building the Robot

Imagine that you are asked to build a robot that will move about a room and manipulate objects. Let's name this robot Immanuel.

The Brain

We can begin to construct Immanuel by starting with a "brain"—typically a PC that will act as the central processing unit. But initially this will be of little use: imagine a computer with no keyboard, no mouse, no screen, no printer, etc.—only a power cord. It can run, but it would be completely isolated. Epistemologically, this is solipsism; psychologically, it is more extreme than autism. Clearly, for the computer to do anything, it must have some ways of getting data (input) and some ways of displaying its results (output). The problem of the first *Critique* is the input problem. The output problem belongs to the second *Critique*.

Transcendental Aesthetic: The Forms of Intuition

If this robot is to be mobile and able to manipulate its environment, one of the first things we need to do is to give it ways of sensing that environment. We might want to begin with a typically human mode of sensing the environment: sight.

In order to give Immanuel vision, we need to hook up a camera to the PC. Imagine that the camera delivers a series of pictures, like frames in a movie, to the central processing unit. If Immanuel just had millions of snapshots cluttering up his memory, he would be totally confused. As the images come in, he needs to process each one of them in two ways. First, he needs to apply a time/date stamp to each image, so that it is possible to figure out which comes before and which comes after. Otherwise he would have no idea of the relationship between any two images. Notice that at this stage (before he tries to communicate with any other robots), Immanuel does not need any more sophisticated notion of time than "before" and "after."

Second, Immanuel needs to put some kind of place or location stamp on each of these images. What would that look like? If Immanuel were stationary, it would be sufficient to stamp each image with the number of degrees of rotation from a central axis (e.g., 37° left or 123° right from the central axis) that the camera has when it is taking the picture. However, if Immanuel is to be mobile, then presumably some system that is less Immanuel-centric would be appropriate. If Immanuel were to stay in a single square room, then his location could be plotted on a simple two or three dimensional grid, depending on whether his elevation could change or not. Today we would probably use a GPS device to provide the coordinates from which the picture was taken as well as the direction in which the camera was pointed, and all of this would eventually be measured in three dimensional spatial coordinates (longitude, latitude, elevation).

The Asymmetry of Space and Time

It is interesting to note that there is an asymmetrical relationship between space and time for the mobile robot. The mobile robot can use a purely internal or agent-centered time system ("before" and "after") without encountering any problems, but it cannot use a purely internal or agent-centered spatial system ("left" and "right") if it is mobile.

Each incoming picture, in other words, needs both a time and place stamp put on it as it comes in—or, as Kant would put it, space and time are necessary forms of intuition. If Immanuel had a picture without these stamps, he would have to discard it simply because he wouldn't know where to put it. It could not count as experience, in Kant's language, until it had a time and place stamp on it. It would not be coherent because, without those stamps, there would be no way that we could determine how it should cohere with the other incoming pictures.

(Obviously, vision won't be the only sense: hearing, smell, touch, and taste may also be added here. Moreover, once we have multiple sources of sensory data, then the need for a common sense [*sensus communis*] becomes evident: all these various sensory inputs need to be stitched together or related to one another. Taken together, in computer language these comprise Immanuel's Basic Input-Output System or BIOS.)

Thinking about building a robot in this way may help students to see more clearly what Kant meant when he said that space and time were necessary forms of intuition. The incoming pictures can be likened to intuitions, and the necessary forms of intuition (space and time) correspond to the way in which each of these incoming pictures must receive a stamp indicating location and time. Thus, to say that space and time are necessary forms of intuition is to say that incoming data must be stamped to indicate the location and time of the incoming data.

Interestingly, we can see in this example of building a robot the priority of time over space, a point that Heidegger among others has made central to his reading of Kant. Clearly, incoming data has to be "stamped" in regard to both space and time, but—as mentioned above—the structure of the spatial stamp changes, depending on whether the robot is stationary or mobile. If the robot is stationary, it needs no sense of place, that is, no sense of its changing location vis-à-vis objects outside of itself. An immobile robot merely needs to note the coordinates of those objects. Once it begins to move, however, it needs to know where it is (the sense of place), and this in turn presupposes that it knows what time it is (i.e., where it is along a temporal continuum, bracketed by "before" and "after.).

Although Immanuel's sense of place is optional, depending on whether he is mobile or not, his sense of time is not. He is always necessarily already in time, already dating things with a "before" and "after." Oddly, there is no temporal mobility in the same sense that there is spatial mobility. Immanuel can move across the room, going from Point A to Point B, or not do so. No such option in available in regard to time. Immanuel simply moves through time at a constant rate which he can neither slow down nor accelerate. He always has a sense of "now" in a way which is different from his sense of "here."

Categories of the Understanding

Of course, when we talk about incoming "pictures," this is in fact very misleading. The term "picture" implies a level of coherence to these data that they don't initially possess. Rather, the incoming data are just a series of pixels or dots of various colours. Let me illustrate the problem in two ways.

First, for those who have used image processing programs (such as Photoshop), they may have encountered the following problem. Imagine telling the program to trace, say, the outline of one person in the photo (so that you can then copy that portion of the photo alone and paste it into another document). Human beings do this quite easily, but computers have an extraordinarily difficult time distinguishing between the figure and the background. All computer programs "see" are pixels, points of colour; they don't see people or meaningful shapes. If there is a sharp difference between the person and the rest of the picture, the computer program can successfully trace it because it can see the differences between the colours. This is not true if there is not a sharp colour difference, for the computer program has no meaningful concept of an object. It only "sees" pixels.

Second, engineers who have put robots on assembly lines have encountered a variant of this same problem. Let's say that the robot's job was hexagonal nuts into two categories: ¼" and 3/8" sizes. The difficulty that the robot faces is that nuts can look quite different, depending on the angle from which they are viewed. This is all the more difficult if the nuts are piled on top of one another. Human beings can distinguish these with comparative ease (and great boredom), but computers have a very difficult time.

Objects, Sameness, and Substance

The first step in solving these kinds of problems is that computers have to be programmed to see the world (i.e., incoming data) in terms of objects. Without the concept of a physical object, the world is utterly chaotic. This has to be written into the computer's software at the most basic level. Incoming data (with its time/place stamp) has to be processed in terms of objects before it can make any sense at all. In other words, very low-level software would have to be written for the robot to insure that it structured incoming data in a particular way, namely, that it looks for continuity over a series of "pictures" by looking for objects that remain the same across time even though we may have different spatial perspectives on those objects so that they don't look identical from one frame to another.

Thinking about these issues in terms of how we would structure a robot's incoming data is helpful in terms of understanding what Kant means by *a priori* and "transcendental." Kant argues that certain concepts are not derived from experience (that is, they are not a posteriori), but rather they are *prior* to experience. For example, the idea that objects persist over time is not something that we learn *from* experience. In fact, sensation (that is, the raw incoming data) simply provides us with an unending stream of data, each of which contains a time/place stamp. Such data only become meaningful when they are organized or structured according to certain rules, such as the notion that there are physical objects that persist over time. This is a notion that we impose on experience, a notion that we use to structure incoming data in order to make those data meaningful. It is prior to experience—in other words, *a priori*—and necessary if we are to structure experience in a meaningful way. It is a necessary condition of the possibility of having meaningful experience at all, and this is what Kant means by "transcendental."

Consider two examples of this. Children sometimes wonder whether the light remains on when the refrigerator door is closed. But we could wonder even more radically about this. Once the door of the refrigerator is closed, do the things inside continue to exist? Of course, we say, but in fact we don't see those objects while the door is closed, and neither would a robot. We have to "teach" (i.e., program) the robot that these objects continue to exist within the refrigerator, even when there is no incoming data that indicate their existence. This programming has to exist before experience rather than be deduced from it.

Consider a second example. Imagine two photos, each of which depicts the same object from a slightly different perspective. However, considered as incoming data, they are very different. Each photo is composed of a number of pixels or dots of colour. Imagine that we were to draw a grid over each photo, dividing it into thousands of tiny squares of colour. Sameness on the level of incoming data occurs when the same pixel within each grid has the same colour. Anytime there is even a small change in the colour, that pixel is no longer the same. We could easily imagine two photos being of the same object but, on the level of pixels, the two would be quite different. The challenge in object-recognition computer programming is to write a set of rules that will allow the computer to determine when two objects are the same and when they are not. The Kantian point about the categories of the understanding here is that these rules have to be in place before experience.

It is also easy to see the incoherence of a world without physical objects. Imagine looking across the room and seeing a book on a desk. Now move a few feet to the side and look at the same desk and book. We automatically stitch these images together in our mind, identifying it as the same book and the same desk, even though both of them now look different because we are viewing them from different angles. Without this automatic work of stitching together these different perspectives into views of the *same* objects, each moment of experience would be unique and incomparable with any other. There must be something that stands underneath this series of pictures and stitches them together, as it were, as the same object. This is the Principle of Permanence of Substance that Kant discusses in the System of All Principles of Pure Understanding in Book II of the Transcendental Analytic.

Cause and Effect

A second necessary category of the understanding is that of cause and effect. "Experience itself—in other words, empirical knowledge of appearances—is thus possible," Kant argues, "only in so far as we subject the succession of appearances, and therefore all alteration, to the law of causality" (B234). Let's consider this in terms of our robot.

As Immanuel "looks" around a room, he receives a series of changing pictures. One of the key distinctions that the robot has to draw is between those changes which occur because of his movement (presuming for the moment that Immanuel, like human beings, can turn his head, move around the room, etc.) and which changes occur because the objects themselves are changing. The concept of substance accomplishes part of this job, but not all of it. We must also have some rule for interpreting these pictures that indicates how one object affects another, a law of cause and effect. Kant's point is that the concept of causal connection is not a concept that we derive from experience; rather, it is a concept that we impose *prior* to experience as a condition of its possibility.

The Categories of the Understanding

We can now see the general outline of the first half of the *Critique of Pure Reason* in terms of our robot analogy. All incoming data must first be given a time and place stamp, that is, they must be subsumed under space and time as forms of intuition. Then they must be organized by certain basic concepts—such as substance and cause-and-effect—in order to be meaningful to us. This second-level of organization involves what Kant calls the categories of the understanding, a set of twelve categories (highly reminiscent of Aristotle's categories) by means of which incoming data (with a time-place stamp) is transformed into meaningful experience.

The transcendental deduction of the categories of the understanding is Kant's attempt to show that these twelve categories are necessary structures of the mind that are imposed prior to experience—indeed, that they have to be imposed on incoming data before those data can count as meaningful experience.

Conclusion: Kant's Transcendental Turn

The robot analogy helps us to understand Kant's transcendental turn more easily. Kant is saying that there are certain basic structures—forms of intuition (space and time) and categories of the understanding (substance, cause-and-effect)—that are necessary structures that the mind must impose on experience for it to be meaningful at all. Kant's point is that these structures are not derived from experience (as the empiricists would have it), but rather are prior to experience (*a priori*) and thus constitutive of it. Yet they are not arbitrary—it is necessary that all incoming data be subjected to these structures—and they do not vary from one individual to the next. Thus they are subjective in one sense (they come from the subject), but they are not subjective in the ordinary sense of that term. They do not vary from one subject to the next. They are necessary structures of experience that come from the subject as necessary conditions of the possibility of experience at all.

It is precisely this combination of subjective (coming from the subject) and necessary that constitutes Kant's transcendental turn, and it is precisely this which we see in building Immanuel. In order for our robot to function, we have to structure its experience according to something very much like the forms of intuition and the categories of the understanding. We may, of course, debate whether Kant's twelve categories are all necessary and whether taken together they comprise the categories that are jointly sufficient for meaningful experience, and such a debate will take us into the heart of the transcendental deduction of the categories of the understanding. At that point, whatever the specific answers are to our questions, it is clear that we have taken Kant's transcendental turn.

CHAPTER TEN

COMPUTATIONAL ABDUCTION: EPISTEMIC MEDIATORS AND THE EXTRA-THEORETICAL DIMENSION OF SCIENTIFIC CREATIVITY

LORENZO MAGNANI
UNIVERSITY OF PAVIA

Computational Modelling and the Enigmatic Concept of Creativity

Peirce's conception of clarity contains the idea that to define the meaning of words and concepts we have to "test," to "subject to" them: the whole conception of some quality lies in its conceivable effects. As he reminds us by the example of the concept of *hardness* "there is absolutely no difference between a hard thing and a soft thing so long as they are not brought to the test" (Peirce 1986). Hence, we can define the "hardness" by looking at those predictable events that occur every time we think of testing some thing.

This methodological criterion can be useful to solve the problem of *creative* reasoning, and to describe, in rational terms, some aspects of the delicate question of a "logic of discovery." What do we mean by "creative," and how can be a "creative process" described?

Much has been said on the problem of *creativity* and hypotheses generation. In the history of philosophy there are at least three important ways for designing the role of hypothesis generation, considered in the perspective of problem solving performances. But all aim at demonstrating that the activity of generating hypotheses is paradoxical, either illusory or obscure, implicit, and not analyzable.

Plato's doctrine of *reminiscence* can be looked at from the point of view of an epistemological argument about the paradoxical concept of "problem-solving": in order to solve a problem one must in some sense already know the answer, there is no real generation of hypotheses, only recollection of them. The activity of Kantian *schematism* is implicit too, resulting from imagination and completely unknowable as regards its ways of working, empty, and devoid of any possibility of being rationally analysed. It is an activity of tacit knowledge, "an art concealed in the depths of the human soul, whose real modes of activity nature is hardly likely ever to allow us to discover, and to have open to our gaze" (Kant 1929, A141-B181: 183). In his turn Polanyi thinks that if all knowledge is explicit and capable of being clearly stated, then we cannot know a problem or look for its solution; if problems nevertheless exist, and discoveries can be made by solving them, we can know things that we cannot express: consequently, the role of so-called *tacit knowledge* "the intimation of something hidden, which we may yet discover" is central (Polanyi 1966).

On the other hand, philosophers of science in the twentieth century, following the *revolutionary* theory developed by Kuhn (1970),[47] have traditionally distinguished between the logic of discovery and the logic of justification. Most have concluded that no logic of discovery exists and, moreover, that a "rational" model of discovery is impossible. In short, scientific creative reasoning should be non rational or irrational and there is no reasoning to hypotheses.

In all these descriptions, the problem is that the definition of concepts like "creativity" and "discovery" is *a priori*. Following Peirce, the definitions of concepts of this sort have not usually rested upon any observed facts, at least not in any great degree; even if sometimes these beliefs are in harmony with natural causes. They have been chiefly adopted because their fundamental propositions seemed "agreeable to reason." That is, we find ourselves inclined to believe them.

Usually this frame leads to a proliferating verbosity, in which theories are often incomprehensible and bring to some foresight just by intuition. But a theory that needs intuition to determine what it predicts has poor explanatory power. It just "makes of inquiry something similar to the development of taste" (Peirce 1986, 254).

A suggestion that can help to solve the enigma of discovery and creativity comes from the "computational turn" developed in the last years. Recent computational philosophy research in the field of cognitive science make use of tools able to give up those puzzling speculative problems, or, at least, to redefine them in a strict pragmatic sense. In fact, taking advantage of modern tools of logic, artificial intelligence, and other cognitive sciences, computational philosophy permits to construct actual models of studied processes. It is an interesting constructive rational alternative that, disregarding the most abstract level of philosophical analysis can offer clear and testable architectures of creative processes.

The development of human society has now reached a technological level in

[47] A perspective established by Reichenbach (1938) and Popper (1959).

which issues concerning the creation and dynamics of information—especially in science—are absolutely crucial. Gradually, philosophical methods and problems are studied and understood in terms of the new information-theoretic notions. Inside the computational philosophy framework, a new paradigm, aimed at unifying the different perspectives and providing some design insights for future ones, rises by emphasizing the significance of the concept of *abduction*, in order to illustrate the problem-solving process and to propose a unified and rational epistemological model of scientific discovery, diagnostic reasoning, and other kinds of creative reasoning (Magnani 2001a). The concept of abduction nicely ties together both issues related to the dynamics of information and its systematic embodiment in segments of various types of knowledge.

Creativity is certainly an important aspect of our definition of "intelligence" but the literature associates many different notions to creativity. This ambiguity has brought to a lack of consensus in the research community. The common views associate to creativity unusual and mysterious qualities that drive the concept of creativity to a confused verbosity. Statements like "to break the rules," "to think different," "to destroy one *Gestalt* in favor of a better one," and "to arrange old elements into a new form," present in the field of psychological research on creativity since 1950s, certainly do not clarify the topic, and seem to lead to the Freudian conclusion that creativity cannot be understood. This conclusion has also been supported by many philosophers who studied conceptual change in science during the second half of the last century. They distinguished between a logic of discovery and a logic of justification (i.e. between the psychological side of creation and the logic argument of proving new discovered ideas by facts). The consequent conclusion was that a logic of discovery (and a *rational* model of discovery) could not exist: scientific conceptual change is cataclysmic and irrational, dramatic, incomprehensible and discontinuous. Many other studies already argued that creativity can be understood (Boden 1991; Sternberg 1999; Sternberg, Kaufman, and Pretz 2002), but paid attention mainly to the psychological and experimental aspects, disregarding the philosophical, logical, and computation ones.

In AI research, however, since Simon, two characteristics seem to be associated to creativity: the *novelty* of the product and the *unconventionality* of the process that leads to the new product. Hence, in a strictly *pragmatic* sense, when we can clarify what behaviour we are looking for, we could implement it in a machine: a methodological criterion enables us to define and consider just those practical effects we conceive to be associated with novelty and unconventionality (cf. Buchanan 2001).

I maintain we can overcome many of the difficulties of creativity studies developing a theory of abduction, in the light of Charles Sanders Peirce's first insights.

The Centrality of Abduction

If we decide to adopt this kind of methodology it is necessary to develop a cognitive model of creativity able to represent not only "novelty" and "unconventionality," but also some features commonly referred to as the entire creative process, such as the expert use of background knowledge and ontology (defining new concepts and searching heuristically among the old ones) and the modelling activity developed in the so called "incubation time" (generating and testing, transformations in the space of the hypotheses). The philosophical concept of *abduction* may be a candidate to solve this problem, and offers an approach to model creative processes of hypotheses generation in a completely explicit and formal way, which can fruitfully integrate the narrowness proper of a merely psychological approach, too experimentally human-oriented.

A hundred years ago, C. S. Peirce (*CP*, 1931-1958) coined the concept of abduction in order to illustrate that the process of scientific discovery is not irrational and that a methodology of discovery is possible. Peirce interpreted abduction essentially as an "inferential" *creative process* of generating a new hypothesis. Abduction has a logical form (fallacious, if we model abduction by using classical logic) distinct from deduction and induction. Reasoning which starts from reasons and looks for consequences is called *deduction*; that which starts from consequences and looks for reasons is called *abduction*.

Abduction is the process of *inferring* certain facts and/or laws and hypotheses that render some sentences plausible, that *explain* or *discover* some (eventually new) phenomenon or observation; it is the process of reasoning in which explanatory hypotheses are formed and evaluated. There are two main epistemological meanings of the word abduction (Magnani 2001a): 1) abduction that only generates "plausible" hypotheses ("selective" or "creative") and 2) abduction considered as inference "to the best explanation," which also evaluates hypotheses. An illustration from the field of medical knowledge is represented by the discovery of a new disease and the manifestations it causes which can be considered as the result of a creative abductive inference. Therefore, "creative" abduction deals with the whole field of the growth of scientific knowledge. This is irrelevant in medical *diagnosis* where instead the task is to "select" from an encyclopedia of pre-stored diagnostic entities. We can call both inferences ampliative, selective and creative, because in both cases the reasoning involved amplifies, or goes beyond, the information incorporated in the premises (Magnani 1992).

Theoretical abduction[48] certainly illustrates much of what is important in creative abductive reasoning, in humans and in computational programs, but fails to account for many cases of explanations occurring in science when the exploitation of environment is crucial. It fails to account for those cases in which there is a kind of "discovering through doing," cases in which new and still unexpressed information is codified by means of manipulations of some external objects (*epistemic mediators*). The concept of *manipulative abduction*[49] captures a large part of scientific thinking where the role of action is central, and where the features of this action are implicit and hard to be elicited: action can provide otherwise unavailable information that enables the agent to solve problems by starting and by performing a suitable abductive process of generation or selection of hypotheses.

The "Internal" Side of Abductive Processes

Sentential Abduction

It is well known that many reasoning conclusions that do not proceed in a deductive manner are *abductive*. What I call *theoretical* (sentential and model-based) abduction (Magnani 1999a) is, from a cognitive point of view, an internal process of reasoning.

Many reasoning conclusions that do not proceed in a deductive manner are *abductive*. For instance, if we see a broken horizontal glass on the floor we might explain this fact by postulating the effect of wind shortly before: this is not certainly a deductive consequence of the glass being broken (a cat may well have been responsible for it). Hence, theoretical abduction is the process of *inferring* certain facts and/or laws and hypotheses that render some sentences plausible, that *explain* or *discover* some (eventually new) phenomenon or observation; it is the process of reasoning in which explanatory hypotheses are formed and evaluated. Moreover, we have to remember that although explanatory hypotheses can be elementary, there are also cases of composite, multipart hypotheses.

First, it is necessary to show the connections between abduction, induction, and deduction and to stress the significance of abduction to illustrate the problem-solving process. I think the example of diagnostic reasoning is an

---

[48] Magnani (2001a, 2002a) introduces the concept of theoretical abduction. He maintains that there are two kinds of theoretical abduction, "sentential'", related to logic and to verbal/symbolic inferences, and "model-based", related to the exploitation of internalized models of diagrams, pictures, etc., cf. below in this paper.

[49] Manipulative abduction and epistemic mediators are introduced and illustrated in Magnani (2001).

excellent way to introduce abduction. I and others (Ramoni et al. 1992) have developed an epistemological model of medical reasoning, called the *Select and Test Model* (ST-MODEL) (Magnani 1992) which can be described in terms of the classical notions of abduction, deduction and induction; it describes the different roles played by such basic inference types in developing various kinds of medical reasoning (diagnosis, therapy planning, monitoring) but can be extended and regarded also as an illustration of scientific theory change. The model is consistent with the Peircian view regarding the various stages of scientific inquiry in terms of "hypothesis" generation, deduction (prediction), and induction.

The type of inference called abduction was studied by Aristotelian syllogistics, as a form of *apāgogē*, and later on by mediaeval re-workers of syllogism. A hundred years ago, Peirce interpreted abduction essentially as an "inferential" *creative process* of generating a new hypothesis. Abduction and induction, viewed together as processes of production and generation of new hypotheses, are sometimes called reduction, that is *apāgogē*.

As I have illustrated above there are two main epistemological meanings of the word abduction: 1) abduction that only generates "plausible" hypotheses (*selective* or *creative*) and 2) abduction considered as *inference to the best explanation*, which also evaluates hypotheses (Figure 1).

To illustrate from the field of medical knowledge, the discovery of a new disease and the manifestations it causes can be considered as the result of a creative abductive inference. Therefore, creative abduction deals with the whole field of the growth of scientific knowledge. This is irrelevant in medical diagnosis where instead the task is to *select* from an encyclopedia of pre-stored diagnostic entities (Ramoni et al. 1992). We can call both inferences ampliative, selective and creative, because in both cases the reasoning involved amplifies, or goes beyond, the information incorporated in the premises.

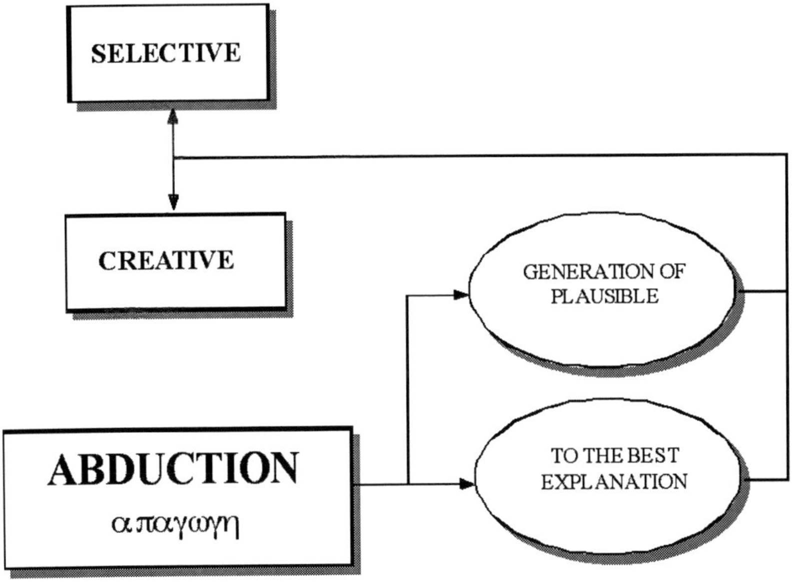

Figure 1. Creative and selective abduction.

Peirce analyses abduction in sentential syllogistic terms, also exploiting the concepts of deduction and induction. To describe how the three inferences operate, it is useful to start with a very simple example dealing with diagnostic reasoning and illustrated (as Peirce initially did), in *syllogistic terms*:

1. If a patient is affected by a pneumonia, his/her level of white blood cells is increased.
2. John is affected by a pneumonia.
3. John's level of white blood cells is increased.
   (This syllogism is known as Barbara).

By deduction we can infer (3) from (1) and (2). Two other syllogisms can be obtained from Barbara if we exchange the conclusion (or Result, in Peircian terms) with either the major premise (the Rule) or the minor premise (the Case): by induction we can go from a finite set of facts, like (2) and (3), to a universally quantified generalisation—also called categorical inductive generalisation, like the piece of hematologic knowledge represented by (1). Starting from knowing—selecting—(1) and "observing" (3) we can infer (2) by performing a selective abduction. The abductive inference rule corresponds to the well-known fallacy called affirming the consequent (simplified to the propositional case).

$$\frac{\varphi \rightarrow \psi \quad \psi}{\varphi}$$

Thus, *selective abduction* is the making of a preliminary guess that introduces a set of plausible diagnostic hypotheses, followed by deduction to explore their consequences, and by induction to test them with available patient data, (1) to increase the likelihood of a hypothesis by noting evidence explained by that one, rather than by competing hypotheses, or (2) to refute all but one.

If during this first cycle new information emerges, hypotheses not previously considered can be suggested and a new cycle takes place. In this case the *nonmonotonic* character of abductive reasoning is clear and arises from the logical unsoundness of the inference rule: it draws defeasible conclusions from incomplete information. All recent logical accounts ("deductive") concerning abduction have pointed out that it is a form of nonmonotonic reasoning. It is important to allow the guessing of explanations for a situation, in order to discount and abandon old hypotheses, so as to enable the tentative adoption of new ones, when new information about the situation makes them no longer the best.

Uncertainty and imperfect information are fundamental characteristics of the knowledge relative to hypothetical reasoning. The nonmonotonic character of the ST-MODEL arises not only from the above mentioned nonmonotonic character of deductive inference type involved in it, but also from the logical unsoundness of the ascending part of the cycle guessing hypotheses to be tested. Finally, we have to remember that in the ST-MODEL the first meaning (see above) of the word abduction is adopted: abduction that only generates "plausible" hypotheses (of course in this case *selective*).

Model-Based Abduction

What exactly is model-based abduction from a philosophical point of view? Peirce stated that all thinking is in signs, and signs can be icons, indices, or symbols. Moreover, all *inference* is a form of sign activity, where the word sign includes "feeling, image, conception, and other representation" (*CP* 5.283), and, in Kantian words, all synthetic forms of cognition. That is, a considerable part of the thinking activity is *model-based.* Of course model-based reasoning acquires its peculiar creative relevance when embedded in abductive processes.

For Peirce a Kantian keyword is synthesis, where the intellect constitutes in its forms and in a harmonic way all the material delivered by the senses. Surely Kant did not consider synthesis as a form of *inference* but, notwithstanding the obvious differences, I think synthesis can be related to the Peircian concept of inference, and, consequently, of abduction. After all, when describing the ways the intellect follows to unify and constitute phenomena through imagination Kant itself makes use of the term *rule* (Kant 1929, A140, B179-180, 182), and also of the term *procedure*. We know that rules and procedures represent the central features of the modern concept of inference.

Most of these forms of "constitution" of phenomena are creative and, moreover, characterized in a model-based way. Let me show some examples of model-based inferences. It is well known the importance Peirce ascribed to diagrammatic thinking, as shown by his discovery of the powerful system of predicate logic based on diagrams or "existential graphs". As we have already stressed, Peirce considers inferential any cognitive activity whatever, not only conscious abstract thought; he also includes perceptual knowledge and subconscious cognitive activity. For instance in subconscious mental activities visual representations play an immediate role.

We should remember, as Peirce noted, that abduction plays a role even in relatively simple visual phenomena. *Visual abduction*, a special form of non-verbal abduction, occurs when hypotheses are instantly derived from a stored series of previous similar experiences. It covers a mental procedure that tapers into a non-inferential one, and falls into the category called "perception". Philosophically, *perception* is viewed by Peirce as a fast and uncontrolled knowledge-production procedure. Perception, in fact, is a vehicle for the instantaneous retrieval of knowledge that was previously structured in our mind through inferential processes. Peirce says: "Abductive inference shades into perceptual judgment without any sharp line of demarcation between them" (Peirce 1955, 304). By perception, knowledge constructions are so instantly reorganized that they become habitual and diffuse and do not need any further testing: "... a fully accepted, simple, and interesting inference tends to obliterate all recognition of the uninteresting and complex premises from which it was derived" (*CP* 7.37). Many visual stimuli—that can be considered the "premises" of the involved abduction—are ambiguous, yet people are adept at imposing order on them: "We readily form such hypotheses as that an obscurely seen face belongs to a friend of ours, because we can thereby explain what has been observed" (Thagard 1988, 53). This kind of image-based hypothesis formation can be considered as a form of *visual* (or *iconic*) *abduction*. Of course such subconscious visual abductions of everyday cognitive behaviour are not of particular importance but we know that in science they may be very significant and lead to interesting new discoveries (Magnani, et al. 1994; Shelley 1996). If perceptions are abductions they are withdrawable, just like the scientific hypotheses abductively found. They are "hypotheses" about data we can accept (sometimes this happens spontaneously) or carefully evaluate.

Peirce gives an interesting example of model-based abduction (Magnani 1999a and 2001a) related to sense activity: "A man can distinguish different textures of cloth by feeling: but not immediately, for he requires to move fingers over the cloth, which shows that he is obliged to compare sensations of one instant with those of another" (*CP* 5.221); this idea surely suggests that abductive movements also have interesting *extra-theoretical* characteristics and that there is a role in abductive reasoning for various kinds of manipulations of external objects (cf. the following section on "action-based, manipulative abduction"). One more example is given by the fact that the perception of tone arises from the activity of the mind only after having noted the rapidity of the vibrations of the sound waves, but the possibility of individuating a tone happens only after having heard several of the sound impulses and after having judged their frequency. Consequently the sensation of pitch is made possible by previous experiences and cognitions stored in memory, so that one oscillation of the air would not produce a tone.

To conclude, all knowing is *inferring* and inferring is not instantaneous, it happens in a process that needs an activity of comparisons involving many kinds of models in a more or less considerable lapse of time. All sensations or perceptions participate in the nature of a unifying hypothesis, that is, in abduction, in the case of emotions too: "Thus the various sounds made by the instruments of the orchestra strike upon the ear, and the result is a peculiar musical emotion, quite distinct from the sounds themselves. This emotion is essentially the same thing as a hypothetic inference, and every hypothetic inference involved the formation of such an emotion" (*CP* 2.643).

Following Nersessian (1999), I use the term "model-based reasoning" to indicate the construction and manipulation of various kinds of representations, not necessarily sentential and/or formal. She proposes the so-called cognitive history and philosophy of science approach, which affords a reframing of the problem of conceptual formation and change in science that not only provides philosophical insights but also pays attention to the practices employed by real human agents in constructing, communicating and replacing representation of a domain. Common examples of model-based reasoning are constructing and manipulating visual representations, thought experiment, analogical reasoning, but also the so-called "tunnel effect" (Cornuéjols et al. 2000), occurring when models are built at the intersection of some operational interpretation domain—with its interpretation capabilities—and a new ill-known domain.

What happens when the abductive reasoning in science is strongly related to extra-theoretical actions and manipulations of "external" objects? When abduction is "action-based" on *external models*? When thinking is "through doing" as illustrated in the simple case above of distinguishing the simple textures of cloth by feeling? To answer these questions I will delineate the features of what I call *manipulative abduction* by showing how we can find in scientific and everyday reasoning methods of constructivity based on external models and actions.

Manipulative Abduction

Manipulative abduction happens when we are thinking *through* doing and not only, in a pragmatic sense, about doing. For instance, when we are creating geometry constructing and manipulating a triangle, like in the case given by Kant in the Transcendental Doctrine of Method (Magnani 2001b). It refers to an extra-theoretical behaviour that aims at creating communicable accounts of new experiences to integrate them into previously existing systems of experimental and linguistic (theoretical) practices. Gooding (1990) refers to this kind of concrete manipulative reasoning when he illustrates the role in science of the so-called "construals" that embody tacit inferences in procedures that are often apparatus and machine based. The embodiment is of course an expert manipulation of objects in a highly constrained experimental environment, and is directed by abductive movements that imply the strategic application of old and new *templates* of behaviour mainly connected with extra-rational components, for instance emotional, esthetical, ethical, and economic.

The hypothetical character of construals is clear: they can be developed to examine or discard further chances, they are provisional creative organization of experience and some of them become in their turn hypothetical *interpretations* of experience, that is more theory-oriented, their reference is gradually stabilized in terms of established observational practices. Step by step the new interpretation—that at the beginning is completely "practice-laden"—relates to more "theoretical" modes of understanding (narrative, visual, diagrammatic, symbolic, conceptual, simulative), closer to the constructive effects of theoretical abduction. When the reference is stabilized the effects of incommensurability with other established observations can become evident. But it is just the construal of certain phenomena that can be shared by the sustainers of rival theories. Gooding (1990) shows how Davy and Faraday could see the same attractive and repulsive actions at work in the phenomena they respectively produced; their discourse and practice as to the role of their construals of phenomena clearly demonstrate they did not inhabit different, incommensurable worlds in some cases. Moreover, the experience is constructed, reconstructed, and distributed across a social network of negotiations among the different scientists by means of construals.

To illustrate this process—from manipulations, to narratives, to possible theoretical models (visual, diagrammatic, symbolic, mathematical)—in a previous work (Magnani 2001a) I have considered some observational techniques and representations made by Faraday, Davy, and Biot concerning Oersted's experiment about electromagnetism. They were able to create consensus because of their conjectural representations that enabled them to resolve phenomena into stable perceptual experiences. Some of these narratives are very interesting.

It is difficult to establish a list of invariant behaviours that are able to describe manipulative abduction in science. As illustrated above, certainly the expert manipulation of objects in a highly constrained experimental environment implies the application of old and new *templates* of behaviour that exhibit some regularities. The activity of building construals is highly conjectural and not immediately explanatory: these templates are hypotheses of behaviour (creative or already cognitively present in the scientist's mind-body system, and sometimes already applied) that abductively enable a kind of epistemic "doing." Hence, some templates of action and manipulation can be *selected* in the set of the ones available and pre-stored, others have to be *created* for the first time to perform the most interesting creative cognitive accomplishments of manipulative abduction.

Moreover, I think that a better understanding of manipulative abduction at the level of scientific experiment could improve our knowledge of induction, and its distinction from abduction: manipulative abduction could be considered as a kind of basis for further meaningful inductive generalisations. Different generated construals can give rise to different inductive generalisations.

Some common features of these tacit templates (Figure 2) that enable us to manipulate things and experiments in science are related to: (1) sensibility towards the aspects of the phenomenon which can be regarded as *curious* or *anomalous*; manipulations have to be able to introduce potential inconsistencies in the received knowledge (Oersted's report of his well-known experiment about electromagnetism is devoted to describe some anomalous aspects that did not depend on any particular theory of the nature of electricity and magnetism; Ampère's construal of experiment on electromagnetism—exploiting an artifactual apparatus to produce a static equilibrium of a suspended helix that clearly shows the role of the "unexpected"); (2) preliminary sensibility towards the *dynamical* character of the phenomenon, and not to entities and their properties, common aim of manipulations is to practically reorder the dynamic sequence of events in a static spatial one that should promote a subsequent bird's-eye view (narrative or visual-diagrammatic); (3) referral to experimental manipulations that exploit *artificial apparatus* to free new possibly stable and repeatable sources of information about hidden knowledge and constraints

(Davy well-known set-up in terms of an artifactual tower of needles showed that magnetization was related to orientation and does not require physical contact). Of course this information is not artificially made by us: the fact that phenomena are made and manipulated does not render them to be idealistically and subjectively determined; (4) various contingent ways of epistemic acting: *looking* from different perspectives, *checking* the different information available, *comparing* subsequent events, *choosing*, *discarding*, *imaging* further manipulations, *re-ordering* and *changing relationships* in the world by implicitly *evaluating* the usefulness of a new order (for instance, to help memory).[50]

The whole activity of manipulation is devoted to build various external *epistemic mediators*[51] that function as an enormous new source of information and knowledge. Therefore, manipulative abduction represents a kind of redistribution of the epistemic and cognitive effort to manage objects and information that cannot be immediately represented or found internally (for example exploiting the resources of visual imagery).[52]

If we see scientific discovery like a kind of opportunistic ability of integrating information from many kinds of simultaneous constraints to produce explanatory hypotheses that account for them all, then manipulative abduction will play the role of eliciting possible hidden constraints by building external suitable experimental structures. So external well-built structures (Biot's construals for example) and their contents in terms of new information and

---

[50] Further aspects of experiment design and its relationship with the problem of communication in science during the transition from the personal to the public domain are given in Gooding and Addis (1999): only a small subset of many observations and measurements performed by individuals of research teams acquire the status of real and public phenomena. Moreover, additional properties of the agent in a scientific experimental setting are described: 1. ability to discriminate between observed results, 2. ability to make judgments about the likelihood of the occurrence of a result, 3. flexibility of the agent's change in perception of the world and his consequent capacity to respond to new information, 4. degrees of competence to build an experiment and observe the results, from novices to experts.

[51] I derive this expression from the cognitive anthropologist Hutchins, that coins the expression "mediating structure" to refer to various external tools that can be built to cognitively help the activity of navigating in modern but also in "primitive" settings. Any written procedure is a simple example of a cognitive "mediating structure" with possible cognitive aims: "Language, cultural knowledge, mental models, arithmetic procedures, and rules of logic are all mediating structures too. So are traffic lights, supermarkets layouts, and the contexts we arrange for one another's behaviour. Mediating structures can be embodied in artifacts, in ideas, in systems of social interactions […]" (Hutchins, 1995: 290-291).

[52] It is difficult to preserve precise spatial relationships using mental imagery, especially when one set of them has to be moved relative to another.

knowledge, will be projected onto internal structures (for instance models, or symbolic and mathematical frameworks) so joining the constructive effect of theoretical abduction. The interplay between manipulative and theoretical abduction consists of a superimposition of internal and external, where the elements of the external structures gain new meanings and relationships to one another, thanks to the constructive explanatory theoretical activity (for instance Faraday's new meanings in terms of curves and lines of force).

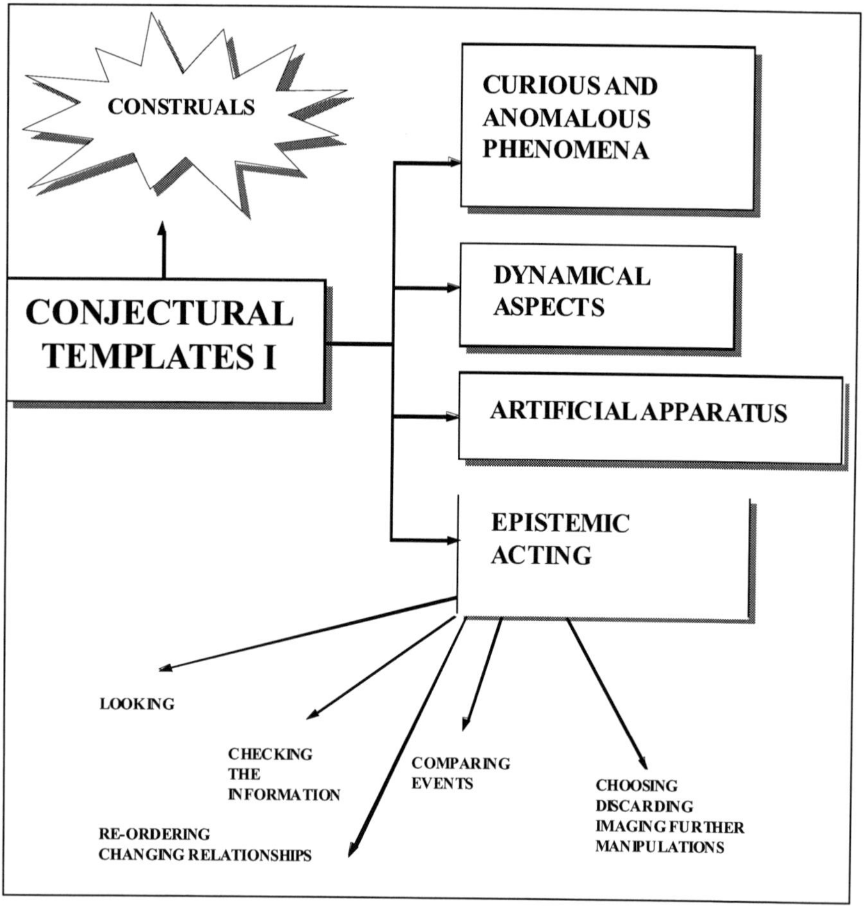

Figure 2. Conjectural templates I

This interplay expresses the fact that both internal and external processes are part of the same epistemic ecology.[53]

Epistemic Mediators

Recent research, taking an explicit ecological approach to the analysis and design of human-machine systems (Kirlik 1998), has shown how expert performers use action in everyday life to create an *external* model of task dynamics that can be used in lieu of an internal model: "Knowing of not, a child shaking a birthday present to guess its contents is dithering, a common human response to perceptually impoverished conditions." Not only a way for moving the world to desirable states, action performs an epistemic and not merely performatory role: people structure their worlds to simplify cognitive tasks but also in presence of incomplete information or of a diminished capacity to act upon the world when they surely have less opportunities to know. *Epistemic action* can also be described as resulting from the exploitation of latent constraint in the human-environment system. This additional constraint grants additional information: in the example of the child shaking a birthday present he "takes actions that will cause variables relating to the contents of the package to covary with perceptible acoustic and kinesthetic variables. Prior to shaking, there is no active constraint between these hidden and overt variables causing them of carry information about each other". Similarly, "one must put a rental car "through its paces" because the constraints active during normal, more reserved driving do not produce the perceptual information necessary to specify the dynamics of the automobile when driven under more forceful conditions " (Kirlik 1998). Moreover, a very interesting experiment is reported concerning short-order cooking at a restaurant grill in Atlanta: the example shows how cooks at various level of expertise use external models in the dynamic structure of the grill surface to get new information otherwise inaccessible.

We know that theoretical abduction certainly illustrates much of what is important in abductive reasoning, especially the objective of selecting and creating a set of hypotheses (diagnoses, causes, hypotheses) that are able to dispense good (preferred) explanations of data (observations), but fails to account for many cases of explanations occurring in science or in everyday reasoning when the exploitation of the environment is crucial. We have illustrated above how the concept of manipulative abduction is devoted to

[53] It is Hutchins (1995: 114) that uses the expression "cognitive ecology" when explaining the role of internal and external cognitive navigation tools. More suggestions on manipulative abduction can be derived by the contributions collected in Morgan and Morrison (1999), dealing with the mediating role of scientific models between theory and the "real world."

capture the role of action in many interesting situations: action provides otherwise unavailable information that enables the agent to solve problems by starting and performing a suitable abductive process of generation or selection of hypotheses.

Also natural phenomena can play the role of external artifactual models: the stars are not artifacts, but under Micronesian navigator's manipulations of the images of them, the stars acquire a structure that "becomes one of the most important structured representational media of the Micronesian system" (Hutchins 1995, 172). The external artifactual models are endowed with functional properties as components of a memory system crossing the boundary between person and environment (for example they are able to transform the tasks involved in allowing simple manipulations that promote further visual inferences at the level of model-based abduction).

Not all epistemic and cognitive mediators are preserved, saved, and improved, as in the case of the ones created by Galileo at the beginning of modern science (see the following subsection). For example, in certain non epistemological everyday emergency situations some skillful mediators are elaborated to face possible dangers, but, because of the rarity of this kind of events, they are not saved and stabilized. Hutchins (1995, 317-251) describes the interesting case of the failure of an electrical device, the gyrocompass, crucial for navigation, and the subsequent creation of substitutive contingent cognitive mediators. These cognitive mediators consist of additional computations, redistributions of cognitive roles, and finally, of the discovery of a new shared mediating artifact in terms of divisions of labour—the so-called modular sum that is able to face the situation.

Finally, we have to observe that many external things that usually are inert from the epistemological point of view can be transformed into epistemic or cognitive mediators. For example we can use our body: we can talk with ourselves, exploiting in this case the self-regulatory character of this action, we can use fingers and hands for counting,[54] we can also use external "tools" like writing, narratives, others persons' information, concrete models and diagrams, various kinds of pertinent artifacts. Hence, not all of the cognitive tools are inside the head, sometimes it is useful to use external objects and structures as epistemic devices. We indicated above that Micronesian navigator's stars, that are natural objects, become very complicated epistemic artifacts, when inserted in the various cognitive manipulations (of seeing them) related to navigation.

[54] Another example is given by the gestures that are also activated in talking, sometimes sequentially, sometimes in an overlapping fashion.

Manipulative Templates

We still know very little about what governs the action-based abduction. At the beginning of this section we already described some central "tacit templates" which characterize manipulative reasoning in science. From the general point of view of everyday situations manipulative abductive reasoning exhibits other very interesting templates (Figure 3). A : 1. action elabourates a *simplification* of the reasoning task and a redistribution of effort across time when we "need to manipulate concrete things in order to understand structures which are otherwise too abstract" (Piaget 1974), or when we are in presence of *redundant* and unmanageable information; 2. action can be useful in presence of *incomplete* or *inconsistent* information—not only from the "perceptual" point of view—or of a diminished capacity to act upon the world: it is used to get more data to restore coherence and to improve deficient knowledge; 3. action as a *control of sense data* illustrates how we can change the position of our body (and/or of the external objects) and how to exploit various kinds of prostheses (Galileo's telescope, technological instruments and interfaces) to get various new kinds of stimulation: action provides some tactile and visual information (e. g., in surgery), otherwise unavailable; 4. action enables us to build *external artifactual models* of task mechanisms instead of the corresponding internal ones, that are adequate to adapt the environment to the agent's needs: experimental manipulations exploit *artificial apparatus* to free new possible stable and repeatable sources of information about hidden knowledge and constraints.

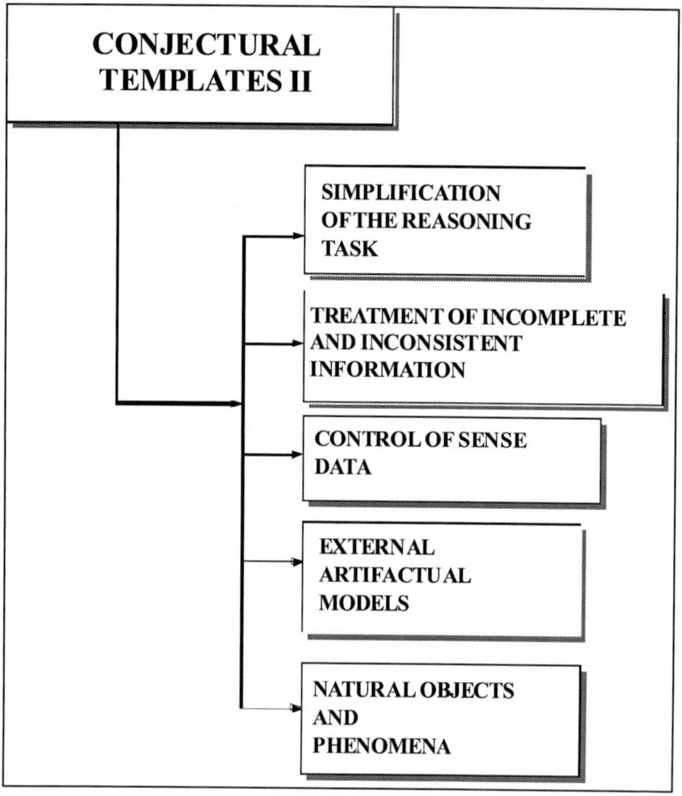

Figure 3. Conjectural templates II

Conclusion

The main thesis of this paper is that abduction is a significant kind of scientific reasoning, helpful in delineating the first principles of a new theory of science. The interdisciplinary character of abduction is central and its fertility in various areas of research evident. The study of the high-level methods of abductive reasoning is situated at the crossroads of philosophy, epistemology, artificial intelligence, cognitive psychology, and logic. The various aspects of abduction I have described provide a better understanding of the processes of explanation and discovery in science. Their analysis certainly increases knowledge about creative and expert inferences that complete the epistemological and cognitive examination of important features of scientific reasoning.

Some research in the area of artificial intelligence has shown that methods for discovery could be found that are computationally adequate for abductively rediscovering—or discovering for the first time—empirical or theoretical laws and theorems.[55] Moreover, the study of diagnostic, visual, spatial, analogical, and temporal reasoning has demonstrated that there are many ways of performing intelligent and creative tasks that cannot be described with only the help of classical logic. However, non-standard logic has shown how we can provide rigorous formal models of many kinds of abductive reasoning such as the ones involved in defeasible, heterogeneous, and uncertain inferences (Magnani, Nersessian, and Pizzi 2002). Abduction is also useful in describing the different roles played by the various kinds of medical reasoning, from the point of view both of human agents and of computational programs that perform medical tasks such as diagnosis (Magnani 2001a, chapter 4). Finally, concrete manipulations of external objects influence the generation of hypotheses: what I call manipulative abduction shows how we can find methods of constructivity in scientific and everyday reasoning based on external models and "epistemic mediators."

I think the cognitive activity of abduction can be further studied in many areas of model-based reasoning (Magnani and Nersessian 2002), such as the ones involving creative, analogical, spatial inferences, and the exploitation of external representations and mediators, both in science and everyday situations, so that this can extend the epistemological and the psychological theory.

References

Boden, Margaret A. 1991. *The Creative Mind: Myths and Mechanisms*. New York, NY: Basic Books.

Buchanan, Bruce G. 2001. Creativity at the metalevel. AAAI-2000 presidential address. *AI Magazine* fall 2001, 13-28.

Cornuéjols, Antoine, Tiberghien, Andrée, and Collet, Gérard. 2000. A new mechanism for transfer between conceptual domains in scientific discovery and education. *Foundations of Science* 5(2): 129-155. Special Issue on- *Based Reasoning in Scientific Discovery: Learning and Discovery*, eds. Lorenzo Magnani, Nancy J. Nersessian, and Paul Thagard. New York, NY: Kluwer Academic/Plenum Publishers.

Gooding, David. 1990. *Experiment and the Making of Meaning*. Dordrecht: Kluwer.

Gooding, David, and Addis, Thomas. R. 1999. A simulation of model-based

[55] On the recent achievements in the area of the machine discovery simulations of abductive creative tasks cf. Magnani, Nersessian, and Pizzi 2002.

reasoning about disparate phenomena. In *Model-Based Reasoning in Scientific Discovery*, eds. Lorenzo Magnani, Nancy J. Nersessian, and Paul Thagard, 103-123. New York, NY: Kluwer Academic/Plenum Publishers.

Holyoak, Keith J. and Thagard, Paul. 1995. *Mental Leaps. Analogy in Creative Thought*. Cambridge, MA: Cambridge.

Hutchins, Edwin. 1995. *Cognition in the Wild.* Cambridge, MA: MIT Press.

Johnson-Laird, Philip N. 1983. *Mental Models: Towards a Cognitive Science of Language, Inference, and Consciousness*. Cambridge, MA: Harvard University Press.

Kant, Immanuel. 1929. *Critique of Pure Reason.* trans. N. Kemp Smith, London: MacMillan.

Kirlik, Alexander. 1998. The ecological expert: acting to create information to guide action. In *Proceedings of the 1998 Conference on Human Interaction with Complex Systems* (HICS'98). Piscataway, NJ: IEEE Press.

Kuhn, Thomas S. 1970. *The Structure of Scientific Revolutions*. 2nd ed. Chicago, IL: University of Chicago Press.

Magnani, Lorenzo. 1992. Abductive reasoning: philosophical and educational perspectives in medicine. In *Advanced Models of Cognition for Medical Training and Practice*, eds. David A. Evans and Vimla L. Patel, 21-41. Berlin: Springer.

—. 1999a. Model-based creative abduction. In *Model-Based Reasoning in Scientific Discovery*, eds. Lorenzo Magnani, Nancy J. Nersessian, and Paul Thagard, 219-238. New York, NY: Kluwer Academic/Plenum Publishers.

—. 2001. *Abduction, Reason, and Science. Processes of Discovery and Explanation.* New York, NY: Kluwer Academic/Plenum Publishers.

—. 2002. Epistemic mediators and model-based discovery in science. In *Model-Based Reasoning in Scientific Discovery,* eds. Lorenzo Magnani and Nancy J. Nersessian, 325-329. New York, NY: Kluwer Academic/Plenum Publishers.

Magnani, Lorenzo, Civita, Sabino, and Previde Massara, Guido. 1994. Visual cognition and cognitive modelling. In *Human and Machine Vision: Analogies and Divergences*, ed. Virginio Cantoni, 229-243. New York, NY: Plenum.

Magnani, Lorenzo and Nersessian, Nancy J., eds. 2002. *Model-Based Reasoning. Science, Technology, Values*. New York, NY: Kluwer Academic/Plenum Publishers.

Magnani, Lorenzo, Nersessian, Nancy J, and Pizzi, Claudio, eds. 2002. *Logical and Computational Aspects of Model-Based Reasoning*. Dordrecht: Kluwer Academic.

Matthews, Michael R., ed. 1989. *The Scientific Background to Modern Philosophy.* Indianapolis, IN: Hackett.

Morgan, Mary S. and Morrison, Margaret, eds. 1999. *Models as Mediators. Perspectives on Natural and Social Science.* Camdridge: Cambridge University Press.

Nersessian, Nancy J. 1998. Kuhn and the cognitive revolution. *Configurations* 6: 87-120.

—. 1999. Model-based reasoning in conceptual change. In *Model-Based Reasoning in Scientific Discovery*, eds. Lorenzo Magnani, Nancy J. Nersessian, and Paul Thagard, 5-22. New York, NY: Kluwer Academic/Plenum Publishers.

Peirce, Charles Sanders. 1955. Perceptual judgments. In *Philosophical Writings of Peirce*, ed. Justus Buchler, 302-305. New York, NY: Dover.

—. (1931-1958) (*CP*). *Collected Papers*, 8 vols., eds. Charles Hartshorne and Paul Weiss (vols. I-VI), and Arthur W. Burks (vols. VII-VIII). Cambridge, MA: Harvard University Press.

—. 1986. *Writings of Charles Sanders Peirce, A Chronological Edition*, vol. 3 (1872-1878). Bloomington, IN: Indiana University Press.

Piaget, Jean. 1974. *Adaptation and Intelligence.* Chicago, IL: University of Chicago Press.

Polanyi, Michael. 1966. *The Tacit Dimension*, London: Routledge and Kegan Paul.

Popper, Karl R. 1959. *The Logic of Scientific Discovery.* London and New Nork: Hutchinson.

Ramoni, Marco, Mario Stefanelli, Lorenzo Magnani, and Gianni Barosi. 1992. An epistemological framework for medical knowledge-based systems. *IEEE Transactions on Systems, Man, and Cybernetics* 22(6): 1361-1375.

Reichenbach, Hans. 1938. *Experience and Prediction.* Chicago, IL: University of Chicago Press.

Shelley, Cameron. 1996. Visual abductive reasoning in archaeology. *Philosophy of Science* 63(2): 278-301.

Sternberg, Robert J., ed. 1999. *Handbook of Creativity.* Cambridge: Cambridge University Press.

Sternberg, Robert J., James C. Kaufman, and Jean E. Pretz, eds. 2002. *The Creativity Conundrum: A Propulsion Model of Kinds of Creative Contributions.* New York, NY: Psychology Press.

Thagard, Paul. 1988. *Computational Philosophy of Science.* Cambridge, MA: MIT Press.

CHAPTER ELEVEN

SENSING AND THINKING THROUGH TECHNOLOGICAL TOOLS: THE IMPACT ON COGNITIVE PROCESSES OF A CHANGE IN THE ORGANIZATION OF THE BOUNDARIES OF BODY. THE CASE OF LAPAROSCOPIC VIDEO-SURGERY

ALBERTO GATTI

DEPARTMENT OF PHILOSOPHY AND SOCIAL SCIENCES, UNIVERSITY OF SIENA

Introduction

Most surgical acts are mediated by external instruments, the surgical ones. These tools fit, on one hand, the surgeon's anatomical and ergonomic features and, on the other hand, the pragmatic function that they have to achieve. Step by step, with the experience, the surgical instruments tend to become something transparent for the surgeon, in the sense that the surgeon becomes so expert in using these instruments, which are, in their turn, shaped so as to fit the humans' physical features, that the instruments almost become part of the surgeon's body itself.

The concept that emerges from this consideration and that has a deep cognitive relevance is the concept of embodiment. Embodiment acquires here a specific meaning. In several contexts within Cognitive Science, the term "embodiment" is used to mean that cognitive processes sometimes take the form of physical actions without going through an isolate abstract theoretical step any more. In this article, I use the term "embodiment" in a different meaning. Embodiment is understood here as the structure of the boundaries of our body. In other words, embodiment refers to the way in which we organize our contact with the environment. We do not have a body that remains stable over time, nor do we always interact with the world only with our naked body, but very often we make use of suitable interfaces that modify the boundaries of our body and, therefore, the kind of interaction between us and the environment.

Alteration of our embodiment has strong consequences both on our sense of our self (Clark 2003) and on the mechanisms of collecting and processing of information. A change in the structure of the boundaries of our body means a change in the structure of the boundaries of our mind.

In this article I take into account laparoscopic video-surgery as a case of embodiment change in surgery. Laparoscopic video-surgery is a kind of mini-invasive surgery in which the surgeon works in a setting and with instruments that are completely different from the setting and the instruments that she encounters in laparotomic surgery. This change in embodiment has important cognitive consequences that I will try to point out.

Main Structure of a Laparoscopic Surgigal Operation

Laparoscopic video-surgery is a mini-invasive surgical technique in which the surgeon intervenes on the anatomical structures contained into the abdomen without opening the abdominal cavity (Kremer, Platzer & Schreiber 1995; Paletto 1997). In this section the main procedures and the main instruments are illustrated that allow such a kind of surgical operation to take place.

Pneumoperitoneum. The first step of every laparoscopic surgical operation is the injection of a gas, which is generally CO_2, into the peritoneal cavity in order to detach the abdominal wall from the viscera and to allow, in this manner, the surgeon to introduce into the abdomen without damage the instruments that are necessary for the operation, to handle such instruments with ease and to observe in a clear way what she is doing throughout the surgical operation.

To execute the pneumoperitoneum, a special needle, called Veress's needle, is introduced in the periumbilical zone or, alternatively, in the left hypochondrium. This needle is connected, by means of a pipe, to a device that emits CO_2 and that is called insufflator. In this manner the abdominal cavity is blown up. The insufflation of CO_2 goes on throughout the surgical operation.

Introduction of the first trocar. Once the pneumoperitoneum has been completed, the next step is to introduce into the abdomen the first trocar. Trocars are special instruments that partially enter the abdomen and that have the function to allow the laparoscopic surgical instruments to be introduced into the abdominal cavity, while preventing, at the same time, CO_2 from going out of the abdomen (Figure 1).

The first trocar is introduced in the periumbilical zone and it is generally put into the abdomen without visual control. If a possibility of damaging internal abdominal structures is present, for example because of the possible presence of adhesions in a patient that has been previously operated on or that shows a particular pathological situation, an alternative, surer, technique can be used for introducing the first trocar. A mini-laparotomy is made in the subumbilical zone and, under visual control, the first trocar is introduced.

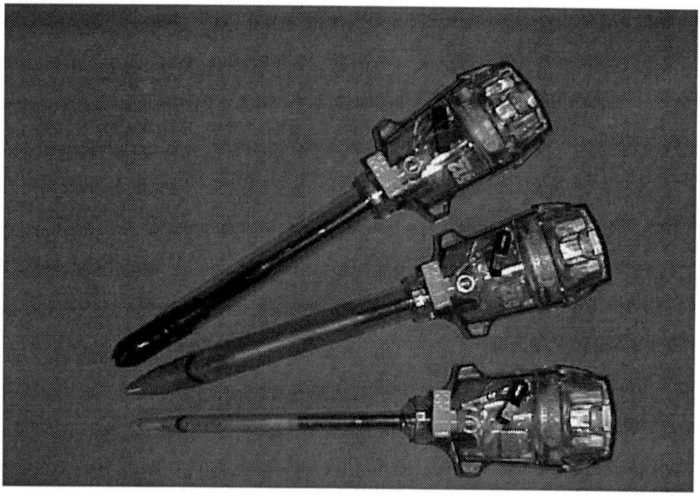

Figure 1. Some kinds of trocar with different diameters. (Taken from Gatti, 2003).

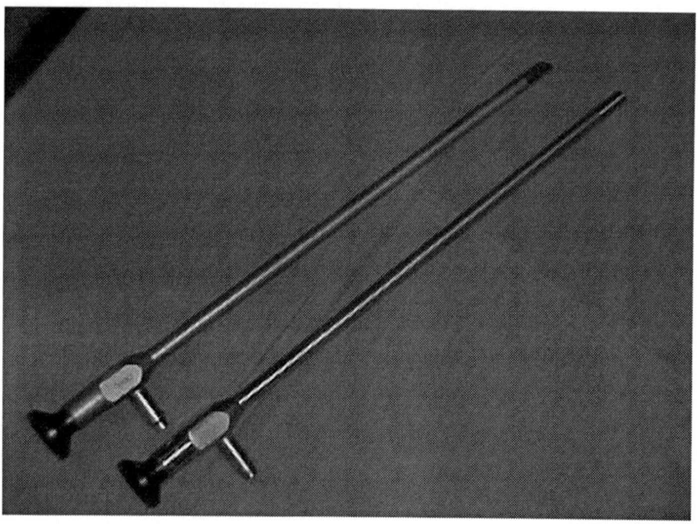

Figure 2. 0° laparoscopic optical probe (below) and 30° laparoscopic optical probe (above). (Picture courtesy of Veronica Gatti).

The first trocar is used to put into the abdominal cavity an optical probe. The laparoscopic optical probes are special devices constituted by a lens system which have the function to visually capture what is going on inside the abdominal cavity during the surgical operation. Laparoscopic optical probes may have a diameter of 5 mm or a diameter of 10 mm. In addition, laparoscopic optical probes differ from each other with respect to the lens angulation: it is possible to distinguish 0°, 25°, 30°, 45° laparoscopic optical probes and also optical probes with a greater lens angulation which reach even a 120° angulation. Two examples of laparoscopic optical probes are shown in fig. 2.

Onto the laparoscopic optical probe a small videocamera is mounted which converts the optical signals into electronic signals and send them, via cable, to a unit that, first, processes and modifies the image and, then, transmits it to a monitor screen. Moreover, a light source sends the light to the laparoscopic optical probe by means of a cable in order to illuminate the peritoneal cavity.

Exploration of the peritoneal cavity After having introduced the laparoscopic optical probe into the abdomen, the surgeon performs an exploration of the various areas of the peritoneal cavity by moving the optical probe and by looking at the monitor in order to diagnostically evaluate the situation, to verify whether pathological signs not previously detected are present and to adjust the plan to follow for the surgical operation.

Introduction of the operating trocars and beginning of the operation The last preliminary step before beginning the therapeutic operation is the introduction into the abdomen of the trocars through which the surgical instruments will pass. The number and the disposition of such trocars will depend on the specific kind of operation that has to be executed. The operating trocars are put into the abdomen under the visual control which is permitted by the artificial visual system.

At this point the laparoscopic surgical operation can begin. Laparoscopic surgical instruments are necessarily different from the instruments used in the traditional laparotomic surgery. The most evident difference is their length. Laparoscopic surgical instruments are longer than the laparotomic ones for the obvious reason that there is, in laparoscopic video-surgery, a greater distance between the surgeon's hands and the operating field due to the fact that the abdomen is not open and that, therefore, the surgeon has to operate from outside. An example of laparoscopic surgical instruments is given in Fig. 3.

Fig. 3 Examples of laparoscopic surgical instruments. The instruments in the picture are used to apply clips in order to close blood vessels. (Taken from Gatti, 2003).

In order to have a clearer idea of the organization of the operating field and of the disposition of the members of the surgical équipe and of the appliances in a laparoscopic surgical operation, Fig. 4 shows one of the possible preparations of the operating field and the consequent disposition of the members of the surgical équipe and of the appliances for the cholecystectomy operation in laparoscopy.

Cognitive Distribution and Embodiment

Laparoscopic video-surgery is a case of complex cognitive distribution. This complexity is due, in particular, to the fact that, besides a cognitive distribution across the members of the surgical équipe, there is a strong cognitive distribution between équipe members and technological appliances. Furthermore, these two kinds of cognitive distribution are not separate, but they tend to be strictly intertwined with each other.

Cognitive distribution across members of the surgical équipe can be viewed under the form of a cooperation in which each member must use a specific piece of knowledge in order to accomplish the surgical operation. The main aspect of the cognitive distribution between members and technological appliances, instead, is constituted by the interaction that the members have with the video system. The surgeon no longer sees the operating field in a direct manner, but she has to observe it on a monitor screen and capture the visual inputs by means of an optical probe endowed with a videocamera.

However, as we have just said, cognitive distribution across members and cognitive distribution between members and technological appliances are deeply intertwined. This can be viewed, in particular, in two aspects that characterize laparoscopic video-surgery.

The first of these two aspects is that, during a laparoscopic surgical operation, all of the members of the surgical équipe have the same view of the operating field thanks to the fact that the image of the operating field is reproduced on a monitor screen and this monitor screen can be equally viewed by all of the équipe members. This is a situation that is not always given in laparotomy, where, many times, only the first operator has a perfect view of the operating field. This means that, in laparoscopic video-surgery, the cooperation across members, that is, the social cognitive distribution that takes place during every surgical operation, is directly supported by the technological system which is devoted to collect visual information from the internal parts of the abdomen, that is, by the cognitive distribution between équipe members and technological appliances.

A

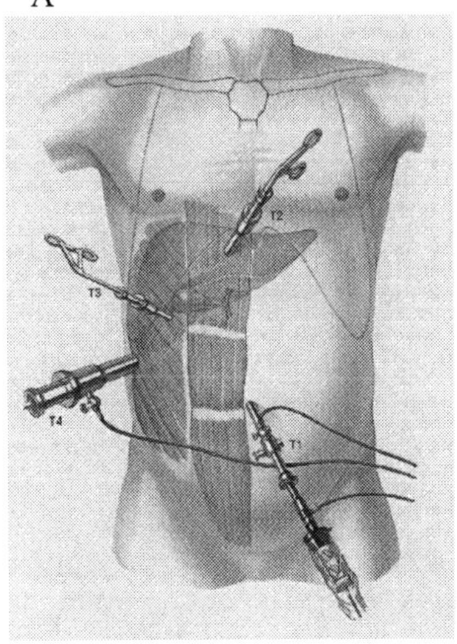

B

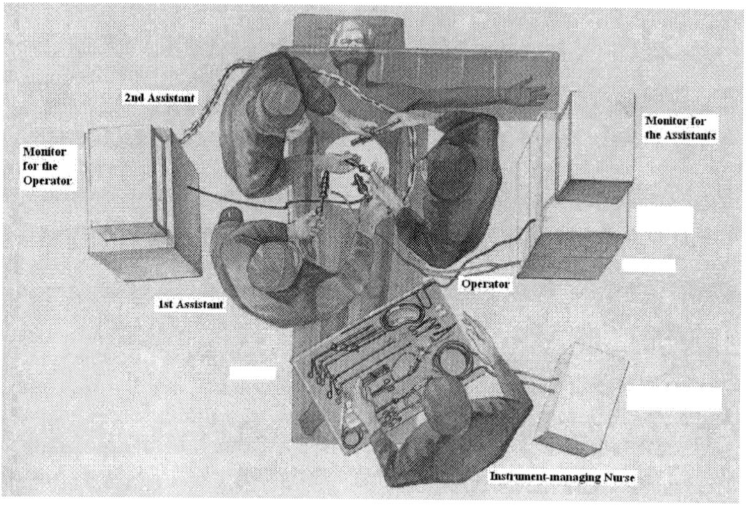

Figure 4. Cholecystectomy operation in laparoscopy.
A: access points. T1: trocar for the optical probe; T2, T3, T4: operating trocars.
B: disposition of the surgical équipe members and of the appliances.
(Taken from Kremer, Platzer & Schreiber, 1995. My adaptation).

The second aspect that points out a deep interaction between cognitive distribution across members and cognitive distribution between members and technological devices is constituted by the way in which the laparoscopic optical probe is generally handled. In most of the laparoscopic surgical operations, the optical probe, which has to be moved in accordance with the surgical actions of the first operator, is handled by a member other than the first operator. This means that not only the cognitive process of collecting visual information is mediated by an external technological device, but the technological device itself is handled by a person other than the cognitive agent that has to use the visual inputs. In other words, we can say that the movements of the first operator's eyes are, in a certain sense, replaced by the movements of the optical probe and that the movements of the optical probe that replace the movements of the first operator's eyes are determined by a person other than the first operator. In this case, the videocamera assistant cooperates with the first operator not with a pragmatic action, but with a visual epistemic action[56]. At the same time, this

[56] Epistemic actions are "actions performed to uncover information that is hidden or hard to compute mentally" (Kirsh & Maglio 1994: 513) and they are distinct from pragmatic

epistemic action is mediated by an external appliance that accomplishes a great part of the visual process.

What we have said so far, together with the technical description given in section 2, points out that laparoscopic video-surgery is one of the cases in which surgery reaches the highest levels of distribution of cognition. Not only there are epistemic actions, not only there is the cognitive use of the operating field as an external representation, not only there is cooperation between équipe members, but there is also a strong cognitive support given by the external video technological appliances. In addition, laparoscopic video-surgery is a kind of surgery that deeply changes the way in which the surgeon has a contact with the operating field. In other words, laparoscopic video-surgery changes the embodiment of the surgeon in her interaction with the operating field. This is determined both by the presence of an artificial video system that replaces a direct vision of the operating field and by the fact that, for the reason that the surgeon acts from outside the abdomen, surgeon's hands are more distant from the operating field with respect to the laparotomic surgery, they cannot be used to touch directly the anatomical structures and they act upon the anatomical structures always by means of longer surgical instruments that reduce the tactile sensibility and can create problems with respect to the sense of distance and to the accuracy of the surgical gesture.

This change in embodiment, that generates a sort of situation of teleoperation, has important consequences on the cognitive processes of collecting information and processing it at the level of the individual surgeon. It is these cognitive consequences that the remainder of this article will try to highlight.

---

actions, which are "actions performed to bring one physically closer to a goal" (Kirsh & Maglio 1994, 513). Within the category of the epistemic actions, I make a further distinction by dividing epistemic actions into *sensorial epistemic actions* and *eliciting epistemic actions*.

A sensorial epistemic action is a structured perceptive action that is performed to explore the environment in a specific way in order to receive from the environment a specific sensation which is used as a representation of information. So, a visual epistemic action, for example, is a sensorial epistemic action. An eliciting epistemic action, instead, is a non-sensorial structured action that is performed to change the configuration of the environment or to stimulate the environment in such a way that it gives previously unavailable information. In other words, a sensorial epistemic action examines a representation by creating it, whereas an eliciting epistemic action creates a representation (a new configuration of the environment or a reaction of the environment) that has to be later examined.

Vision

It can be said that one of the targets of laparoscopic video-surgery is to give rise to a mini-invasive surgery by means of technological appliances that reproduce an operating situation which be, at the sensation level, as similar as possible to the operating situation that the surgeon encounters in laparotomic surgery. From this point of view, the appliances that support a laparoscopic surgical operation would aim at becoming “transparent technologies” (Clark 2003), that is, tools that are so well integrated with our own lives, with our ways to confront problems, with our ways to interact with the rest of the world that they become almost invisible while we use them (Norman 1999).

It cannot be ignored that this aspect of transparency of technological devices that try to give the surgeon the impression to operate in a situation similar to the laparotomic one is actually present in laparoscopic video-surgery and it is due, especially, to the video system that allows the surgeon to see the results of her own actions. The surgeon acts upon real anatomical structures by means of the surgical instruments, but she observes what she is doing on the monitor screen and this certainly constitutes a change in embodiment that creates a completely different physical behaviour in the surgeon. But, if the surgeon, thanks to the images of the video system, can have a good control upon her own gestures and, consequently, upon the anatomical structures, she is likely to find herself in a situation similar, at the level of sensation, to the one that she encounters in laparotomy.

However, the transparency of the technological devices used in laparoscopy is not complete. The video system, which is the instrument that allows the surgeon to maintain a control of the situation similar to the one that she has in laparotomic operations, is also the element that creates difficulties as to the control of the gestures involved in a surgical operation. The main problem that can be found in laparoscopy is a gap between surgical gesture and vision: the image on the monitor screen is managed by a videocamera and, for this reason, such image gives sensations of dimension and depth that not always correspond to the real situation and that may create problems in the adjustment of the surgical gestures. In other words, the surgeon acts on the basis of the images on the monitor screen; these images visually reproduce the operating field with dimensions that are different from the real ones, but the surgeon is physically acting upon the real anatomical structures and this causes difficulties in adapting the surgical gestures to the real anatomical situation. It is necessary for the surgeon a repeated training in order to adapt herself to this new embodiment in which she has to bring into coordination actions performed upon anatomical structures that she does not see directly with visual information about those anatomical structures observed on a monitor screen. After adequate training, however, this gap between gesture and vision is overcome and the coordination between visual inputs and gesture tends to become very similar to the one that takes place in laparotomy.

What we have said so far in this section is summarized in some assertions made by the surgeon that was my first informant during my in-the-field research and that I call here "The Fox". The assertions I am referring to are the following ones:

> *Me: Quali sono le differenze tra un intervento in laparoscopia e un intervento in laparotomia?*
> *[...].*
> *"The Fox": [...]le differenze sostanzialmente sono quelle di porre un filtro tra sé, la propria visione, e l''oggetto e questo filtro quanto maggiormente è connaturato con il gesto, tanto minore sarà la differenziazione rifrattiva che ci può essere[...], nel senso che noi abbiamo una differenza tra quello che vediamo e il gesto che facciamo, ma[...]meglio si completa questa sorta di salto di visione e gesto, tanto più alla fine diventerà naturale e quasi normale[...].*
> *Me: Quasi come una laparotomia, a quel punto?*
> *"The Fox": Quasi come una laparotomia.*
>
> Me: What are the differences between an operation in laparoscopy and an operation in laparotomy?
> [...].
> "The Fox": [...]the differences are basically the ones of putting a filter between oneself, one's vision, and the object and the more connatural this filter is with the gesture, the smaller the refractive differentiation will be that can take place[...], in the sense that we have a difference between what we see and the gesture that we do, but[...]the better this sort of gap between vision and gesture is bridged, the more natural and almost normal it will become in the end[...].
> Me: Almost like a laparotomy, at that point?
> "The Fox": Almost like a laparotomy.

So far we have dealt with the relationship between mediated visual inputs and surgical gestures in laparoscopy and with the transparent aspect that laparoscopic appliances may exhibit. Now I want to concentrate on the specific and important cognitive changes that the new embodiment constituted by laparoscopic devices generates at the level of visual processes.

The optical probe that is used in laparoscopic video-surgery allows the surgeon to execute a visual exploration of the abdominal cavity which is more accurate than the visual exploration that can be performed in the case of a laparotomic operation. This is due to the fact that the laparoscopic optical probe can explore abdominal zones that the human eye cannot reach or cannot see very well, it can arrive nearer the anatomical structures than the human eye, it may offer different vision angulations according to the type of lens, allowing the surgeon, this way, to have a wider view of the abdominal cavity and to explore hidden areas; finally, the videocamera mounted on the optical probe allows the image to be enlarged by means of the zoom system.

From a cognitive point of view, the new embodiment constituted by the laparoscopic optical probe and, in general, by the video system used in laparoscopy changes and improves the surgeon's visual epistemic actions and, consequently, the visual representations that the surgeon collects in her own interaction with the operating field. In laparoscopic video-surgery, surgeon obtains new, clearer visual representations of anatomical structures that she had not been able to see well in laparotomy or that she had not seen at all and she had only been able to touch. A strict relationship between epistemic actions and representations is evident here: an improvement in visual epistemic actions generated by an external artifact brings to an improvement in the visual representations of the same identical information.

A demonstration of this fact and, in addition, of the fact that different representations of the same information may bring a cognitive agent to different cognitive processes and to different decisions about what to do in a particular situation is that, in some cases, it happened that some surgical mistakes were made in laparoscopy because of the fact that the operator could see in a clear way anatomical structures that could not be seen very well in the laparotomic version of the same operation or that could only be felt by touch. For this reason, the operator could not recognize those anatomical structures in the laparoscopic surgical operation and this was at the basis of the surgical mistakes:

> *"The Fox": [...]dove la videochirurgia si è maggiormente diffusa è proprio in quei tipi di interventi in cui la visione videoendoscopica è migliore di quella aperta.[...].[...]il successo che ha avuto la colecistectomia laparoscopica è perché si vede bene, molto bene, la si vede meglio. Cioè, tu riesci a vedere strutture che normalmente intravedi o senti con le dita, al punto tale che ci sono stati anche degli errori di tipo chirurgico nei confronti della colecistectomia, perché si riusciva a vedere[...]delle strutture talmente bene che si è andati a legare determinate strutture che non dovevano essere legate: quando sono stati legati e tagliati, sezionati dei coledoci[...]perché[...]tu li vedi, pur vedendoli non li riconosci in termini di abitudine, di esperienza e si son legati.*

> "The Fox": [...]where video-surgery has mostly spread it is in those kinds of operations in which the video-endoscopic vision is better than the open one.[...].[...]the success that laparoscopic cholecystectomy has obtained is because you see it well, very well, you see it better. That is, you can see structures of which you usually catch a glimpse or that you feel with your fingers, to the point that there were even mistakes of a surgical type with respect to the cholecystectomy, because you could see[...]certain structures so well that

certain structures were bound that didn't have to be bound: when common bile ducts were bound and cut, sectioned[...]because[...]you see them, even if you see them you don'trecognize them in terms of habit, of experience and they were bound.

So, such mistakes are explained by a contrast between an internalized representation of anatomical structures that has been generated through the experience in a laparotomic context and a new visual representation of the same anatomical structures that is encountered in a laparoscopic context. This contrast makes the mechanism of recognition difficult or impossible and may bring the surgeon to make mistakes.

From the point of view of vision, in laparoscopic video-surgery the surgeon becomes an augmented cognitive agent thanks to the presence of an external video system that modifies and improves her capacity of capturing visual inputs. Laparoscopic video-surgery, in this sense, demonstrates that a change in epistemic actions (in this case, sensorial epistemic actions) causes a change in the representations used in cognitive processes and that human cognitive agents find themselves in front of different representations of the same information and, in addition, it demonstrates that the boundaries of our mind are deeply intertwined with the boundaries of our body and that the boundaries of our body are flexible.

Touch

One of the most important consequences of the setting in which the surgeon finds herself to operate in laparoscopic video-surgery is that she no longer can use her own "naked" hand to touch the anatomical structures. Direct palpation, which has a crucial importance in traditional laparotomic surgery, disappears in laparoscopic technique. It is vision, an improved vision, as we have seen, that, in laparoscopic video-surgery, holds the main role at a diagnostic level. The surgeon mostly uses a visual semeiology:

"The Fox": [...]sensazioni tattili ci possono ancora e ci devono essere; sicuramente hanno meno spazio di prima, cioè, adesso[...]hai una semeiotica visiva, cioè, tu vedi e decidi.

"The Fox": [...]there can still be and there must be tactile sensations; they certainly have less space than before, that is, now[...]you have a visual semeiology, that is, you see and decide.

Notwithstanding this, palpation at a diagnostic level is still present in laparoscopy, but it is mediated by special instruments, that is, metal bars that are introduced into the abdomen through a trocar and are pushed against anatomical structures in order to evaluate their consistence or are used to touch behind an anatomical structure in order to test the presence or the absence of some elements that cannot be seen.

Once again, we are in presence of a change in embodiment that causes a change in sensorial epistemic actions and, consequently, in the sensorial representations that are used in cognitive processes. In laparoscopic video-surgery there are, at a diagnostic level, new tactile representations generated by new tactile epistemic actions. In particular, these new tactile representations are less accurate than the ones that can be obtained in laparotomic surgery and, for this reason, as we have already said, vision acquires a greater importance.

Besides tactile gestures and tactile sensations that have a specific diagnostic value, another important element are the tactile sensations that the surgeon feels while performing surgical actions during a surgical operation. In laparoscopic video-surgery there is a deep change also for this kind of tactile sensations. The impact with the anatomical structures is different from the one that the surgeon has in laparotomic surgery and this is due to the fact that the surgeon uses longer instruments that drastically reduce the tactile sensation that is generated by the contact between surgeon and anatomical structures. This change in embodiment that causes a reduction in tactile sensation has important consequences on the sense of control. The surgeon sees what she is doing through the monitor screen, but not always she has a tactile sensation of what she is doing, not always she physically realizes that she is touching anatomical structures and this can reduce the sense of control that the surgeon feels.

On the basis of "The Fox's" personal communications collected through interviews, I have found that it is possible to distinguish two situations of tactile sensation, on the part of the surgeon, of what the surgeon herself is doing during a laparoscopic surgical operation. The point of departure of my analysis are these important statements made by "The Fox":

> *[...]mi sono accorto che, nel momento in cui ti viene il gesto consueto e fai la determinata cosa che hai imparato a fare magari, invece, in gesto tradizionale, il tuo salto di qualità, il salto di qualità, a mio modo di vedere, lo riesci ad acquisire nel momento in cui tu, oltre a vedere, che ha la sua importanza e vedi quello che fai e quindi puoi correggere il gesto, hai anche coi ferri la sensazione di quello che stai facendo.[...]non in tutti i gesti tu riesci a capire che stai toccando i tessuti, ma in certi sì[...]. Non è sempre facile; certe volte hai una sensazione talmente leggera che non hai...tu la vedi (la struttura anatomica) e ti sembra di sentirla, però ci son certi momenti in cui, in effetti, senti anche quello che stai facendo e allora lì nasce la capacità[...]di impattarti con i tessuti che stai trattando, quindi che non deve essere soltanto un virtuale che tu vedi, se possibile, ma anche una sensazione reale[...].*

> [...]I have realized that, at the moment in which you do the usual gesture and you do the thing that, perhaps, you have learned to do, instead, with the traditional gesture, you can acquire your jump in quality, the jump in quality, in my opinion, at the moment in which, besides seeing, which has its own importance and you see what you do and, therefore, you can correct the gesture, *you also have, with the surgical instruments, the sensation of what you are doing*.[...]not in all of the gestures you can understand that you are touching the tissues, but in certain gestures you can[...]. It is not always easy; some times you have such a light sensation that you don'thave...*you see it* (the anatomical structure) *and it seems to you to feel it*, but there are certain moments in which *you indeed feel too what you are doing* and then, at that moment, the capacity arises[...]of having an impact with the tissues you are handling, which does not have to be, thus, only a virtual that you see, if possible, but also a real sensation[...].

On the basis of these statements, I analyse from a cognitive point of view the two situations that emerge from "The Fox's" words.

The first situation is the one in which the surgeon has the impression of feeling an anatomical structure because she sees on the monitor screen her own surgical instrument touching that anatomical structure. This can be defined as a virtual tactile sensation induced by a visual representation: the visual representation on the screen shows the image of a physical contact and the surgeon has the impression of feeling that physical contact. This is an interesting case from a cognitive point of view, because the surgeon uses only one real sensation, the one given by the external visual representation and, on the basis of the specific features of this sensation, creates a tactile representation that does not correspond to any real tactile stimulation. It can be said that the surgeon cognitively works with two completely different representations, one constituted by a sensation generated by a contact with a particular stimulus and the other internally constructed with a sort of synesthetic process on the basis of the features of the sensation that constitutes the first representation. We can assume that the second representation, that is, the tactile representation, is constituted by tactile components that the surgeon has internalized through the experience in laparotomic surgical operations. The virtual tactile sensation is used to reinforce the control on what the surgeon is doing.

The second situation is the one in which the surgeon has not an impression of feeling an anatomical structure she is touching, but she really feels that anatomical structure. In this case the tactile representation is not constructed on the basis of a visual representation, but it is generated by a tactile stimulus and, in particular, such tactile representation is no longer constituted by internalized tactile sensations, but by a new tactile sensation, in which the sensation is mediated by a surgical instrument which is longer than the laparotomic ones. The surgeon is using, in this case, a new tactile representational code, she is experiencing a new tactile impact with the anatomical structures. The new tactile impact and the modified visual contact with the operating field that we have considered above are the main elements that result from the new embodiment created by laparoscopic video-surgery.

Conclusion

Laparoscopic video-surgery demonstrates, first of all, that the boundaries of the human cognitive system are not confined to the individual human agent, but they include external technological artifacts that perform cognitive processes and that have the function of external cognitive scaffolding. In addition, and this is the point of the article, laparoscopic video-surgery demonstrates that the structure of the cognitive processes is deeply dependent on the way we organize our perceptive contact with the environment. The technological artifacts used in laparoscopic video-surgery are external elements that modify the surgeon's perceptive relationship with the patient's body and that, in this manner, modify the sensorial inputs that the surgeon uses as representations of information.

To change the perceptive structure of our body sometimes means to change the structure of our mind.

References

Clark, A. 2003. *Natural-born Cyborgs: Minds, Technologies, and the Future of Human Intelligence.* New York: Oxford University Press.

Gatti, V. 2003. La laparoscopia nell'urgenza chirurgica addominale. Specialization dissertation, School of Specialization in General Surgery II-Emergency Surgery, University of Pavia, Pavia, Italy.

Kirsh, D., and P. Maglio, P. 1994. On distinguishing epistemic from pragmatic action. *Cognitive Science* 18: 513-549.

Kremer, K., W. Platzer, H. W. Schreiber, eds. 1995. *Chirurgische Operationslehre Band 7/2. Minimal-invasive Chirurgie: Video-laparoskopische und video-thorakoskopische Chirurgie.* Stuttgart: Georg Thieme Verlag. English Translation by J. Grossman. 1999. *Atlas of Operative Surgery: Surgical Anatomy, Indications, Techniques, Complications. Minimally Invasive Abdominal Surgery,* eds. K. Kremer, W. Platzer, H. W. Schreiber, and F. M. Steichen. New York: Thieme Medical Publishers.

Norman, D. A. 1999. *The Invisible Computer.* Cambridge, MA: MIT Press.

Paletto, A. E. ed. 1997. *Nuovo trattato di tecnica chirurgica (Vol. 2). Pareti addominali, stomaco, duodeno, digiuno, ileo.* Torino: UTET.

CHAPTER TWELVE

GROWING MINDS, COMPUTABILITY, AND THE POTENTIALLY INFINITE

DARREN ABRAMSON
DEPARTMENT OF PHILOSOPHY, DALHOUSIE UNIVERSITY

Introduction

In this paper, I introduce an intuitive notion of sets which are neither finite nor infinite. Then, I make this notion precise by defining what I call potentially infinite sets. I argue that this notion is helpful in understanding properties of human minds, and—in a related fashion—can be usefully applied to notions of computability. It is useful to start by noting that the following inference is fallacious.

Set T is not finite. Therefore it is infinite.

Similarly, it is fallacious to conclude that because a set is not infinite, that it is finite. It is neither a logical truth nor a theorem of classical set theory that an arbitrary set is either finite or infinite.[57]

Some objects seem well suited to being placed in sets; for example, it seems unproblematic to discuss sets of material objects, properties, and of course numbers. Consider the set of words I am in practice capable of understanding. Is this set finite or infinite? I claim that neither is the case. For, suppose that the set of these words is finite. It would seem that we could arbitrarily conjoin words as long as we wanted to and make ever more words.

On the other hand, suppose that this set of words is infinite. This too is counterintuitive. It seems to grant too much to our powers to think that, if confronted with any one of an infinite number of words we could understand each one. Suppose that to understand each word a minimal amount of experience is both a necessary and sufficient condition.

[57] For relevant results, see Ellentuck 1965.

Now consider the set of all such experiences. We seem to have to grant both that it is infinite and deny that any person can have experienced every member of it. There just isn't ever enough time. So, while there is no fixed bound on the number of words a person can understand, no one at a given time can claim to understand an infinite number of words. To say that the set of words I am capable of understanding is finite is false, and to say that it is infinite likewise seems mistaken.

Let us look at another set which we want to say is neither finite nor infinite. How many days would you like to live? Bernard Williams in The Makropulos Case concludes that

> . . . an eternal life would be unliveable. In part . . . that is because categorical desire will go away from it: in those versions . . . in which I am recognizably myself, I would eventually have had altogether of myself. But equally, at times earlier than that moment, there is reason for not dying. Necessarily, it tends to be either too early or too late. . . . If that is any sort of dilemma, it can, as things still are and if one is exceptionally luck, be resolved, not by doing anything, but just by dying shortly before the horrors of not doing so become evident.(Williams 1976, 100)

What would it mean to die just before one realizes that by not doing so, one will have had an infinitely long life? Williams is guilty of the fallacy I just mentioned. Either one can wish for a finite life of just the right length—shortly before the horrors become evident. Or, Williams tells us, one is doomed to the self-destroying boredom of an infinite lifespan. This forced choice is an example of the false dichotomy of finite and infinite sets.

Another argument which fallaciously assumes "all sets are finite or infinite" is one of Putnam's against functionalism. He asks us to reflect on the possibilities intrinsic to human experience and discourse. No completed semantics, says Putnam, can be provided or the formal language which he argues the functionalist must believe underlies our mental life. Putnam says, "Experience tells us that no human society is unsurpassable.

For any human society, there is a possible other society which is more sophisticated, which has modes of conceptualizing and describing things which members of the first society cannot understand without years of specialized study" (Putnam 1991, 89). The lesson which Putnam wants to draw is that no one can have a completed theory of natural language use. He has established that no finite theory will do—this follows from the surpassability of human societies. However, this does not mean that being capable of understanding a final theory is tantamount to ". . . [surveying] all modes of human linguistic existence—including those that will transcend his own" (ibid). The functionalist can deny that understanding the theory of natural language involves reflecting on infinitely many rules. They can claim that such understanding is the possession of a finite set of rules which can be arbitrarily extended in possible contexts of human experience.

To run a distance of 1 unit, how many total distances must Achilles run? Zeno's paradox involves the observation that on the one hand, Achilles must run an infinite number of distances (1/2, 1/4, 1/8, etc.). On the other hand, despite Achilles' great strength and speed, he can only accomplish a finite number of tasks in the finite amount of time it takes him to run a distance of 1 unit. We may imagine that Achilles truly has no limit on his strength and speed. So, if we give him any number of flags, he can set them down at the halfpoints of the remaining distances. However, we still feel certain that it is impossible to take an infinite number of flags and set them all down by the time he has completed his run.

In each case, I want to preserve our idea that neither finitude nor infinitude fits the bill. In each case we have a quantity which cannot be bounded by a finite number. However, we resist the idea that the quantity is infinite in size. Let us call a quantity which is neither finite nor infinite "potentially infinite." Given a set which is potentially infinite, we will have only traversed a finite portion of it. However, it is always possible to traverse more of the given set. Yet at no point can an infinite amount of the set be traversed.

Potentially Infinite Sets

Now we will give a precise definition to potentially infinite (PI) sets, sets which are neither finite nor infinite, and look at a structure which satisfies this definition.

A set S of natural numbers is PI =df

(a) $\neg\exists x\Box\forall y(Sy \rightarrow y \prec x)$ (S cannot be finite), and

(b) $\Box\neg\forall x\exists y(Sy \wedge y \succ x)$ (S cannot be infinite).
where x and y range over N.

The first clause says "there does not exist an x such that necessarily for all y if y is in the set, then y is less than x." It captures the intuition that if someone offers a fixed finite bound on the number of objects in a potentially infinite set, then there is a possible world in which there are more objects in the set than that bound. The second clause says that in no possible world is it the case that the number of objects in the set gets to be infinite.

Now we will look at a potentially infinite set. Consider possible worlds which are labeled by the natural numbers to give us world1, world2, and so on. Let "less than" be the accessibility relation among possible worlds, so that in world47, world48 is accessible, for example. Now, suppose that set S has the following modal membership conditions. n is in S in world m just in case n < m. S so defined is PI. To see this, let us look at S in world k. Assume for reductio that S does not satisfy the first clause of our definition. Then there is some n such that is necessarily bigger than any m in S. However, it is possible that k + n is in S, and k + n is greater than k. So we have a contradiction, and therefore there is no such n. S satisfies the second clause of our definition. Again, fix world k. Then k+1 is greater than any element in S, and so S is not infinite. Since world k was arbitrarily chosen, failure to be infinite holds necessarily.

The notion of a potentially infinite set has its roots in Aristotle's idea of potential infinities. A.W. Moore shows how Aristotle applies his distinction between potential and actual infinities.

> [Given Aristotle's distinction], "This body is infinitely divisible" can be glossed as either (1) For every natural number n, there is a possible situations, such that this body is divided into more than n parts in s or (2) There is a possible situation s, such that for every natural number n, this body is divided into more than n parts in s. . . . Aristotle would have accepted (1), and he would have found (2) unintelligible. It was thus that he solved the paradoxes of the infinitely small. (Moore 2001, 42)

Aristotle's account of infinite divisibility thus satisfies the two clauses of the definition of a potentially infinite set. Potentially infinite sets as introduced here require a modal context. Aristotle's semantics for the modality in potential infinities interprets as quantifying over all future moments, and as quantifying over some future moment or moments. This means that for any n, there is a moment in time in which a potentially infinitely divisible body will in fact be divided into more than n parts. The present account allows other semantics, including possible worlds which make no reference to time.

Let us see that by considering each of the sets we began with as potentially infinite we are able to satisfy our intuitions. First, will desiring a potentially infinite number of days to live assuage Williams' worries? On the one hand, this scenario would mean that on any given day, the number of days I have lived is smaller than some number. However, at the same time, it is possible that there are more remaining. Interpreting Sa as "a is an index on a day I live", the first clause becomes "there does not exist an x such that necessarily for all y, if I live to enjoy y many days, then y is less than x. The second clause becomes "it is necessarily not the case that for all x there is some y such that I live y many days, and y is greater than x". As for Achilles, we suppose that his capacity for placing flags is without bound. Can he place an infinite number of flags down? Certainly not. Is there a finite bound on the number of flags he can place? No. However, given any amount of flags he has placed during his run, it is always possible that he could have trained just a little bit harder to place one more flag before finishing. In short, there is no bound for the possible number of flags to be placed, and we can't ever say that for any natural number, there have been more flags placed than that number.

Finally, consider the number of words a person can understand. There is no number which limits the number of words they can comprehend. However, at no point can they be said to have understood an infinite number of words.

Computability and the Potentially Infinite

Turing: What Can People Compute?

Of special interest is an instance of characterizing human capacity which fallaciously assuming that sets are either finite or infinite. Alan Turing was interested in asking the question "What functions over the natural numbers can a person compute?" Human computability is nothing more, Turing argued, than the ability to recognize and write a finite number of symbols on paper, modify those symbols and change one's state by following rules determined by the current state of mind and the symbols observed. Here is a key portion of Turing's analysis.

> The behaviour of the [human] computer at any moment is determined by the symbols which he is observing, and his "state of mind" at that moment. We may suppose that there is a bound B to the number of symbols or squares which the computer can observe at one moment. If he wishes to observe more, he must use successive observations. We will also suppose that the number of states of mind which need be taken into account is finite. The reasons for this are of the same character as those which restrict the number of symbols. If we admitted an infinity of states of mind, some of them will be "arbitrarily close" and will be confused. (Turing, 1936) (pg. 137 in Martin Davis, 1965).

The last portion of this argument is fallacious. Gödel noted this.

> Turing . . . gives an argument which is supposed to show that mental procedures cannot carry any farther than mechanical procedures. However, this argument is inconclusive, because it depends on the supposition that a finite mind is capable of only a finite number of distinguishable states. What Turing disregards completely is the the fact that mind, in its use, is not static, but constantly developing . . . although, at each stage of the mind's development the number of its possible states is finite, there is no reason why this number should not converge to infinity in the course of its development.(Wang 1974: 325)

Gödel identified informally a concept that we have made precise, a set of determinate size which is neither finite, nor infinite. His comments agree with both clauses of the definition of a potentially infinite set. At any time the number of possible states is finite, and there is no bound on how many states it is possible for the mind to be in. The first clause reveals Turing's error. From the assumption that the number of states a human mind can be in is not finite, it does not follow that at any given moment that mind can be in an infinite number of states. And, if at no point can a human mind be in an ininite number of states, then it is besides the point whether an infinite number of states would be arbitrarily close and confused.

Gödel's comments have been dismissed. For example, Judson Webb (Webb 1980) argues that even if mind is capable of infinitely many states in its final development, this tells us nothing about human computation procedures. It remains to be shown that we could make effective use of all these states. He concludes that an infinite mind doesn't get us any where near the construction of an effective, non-Turing computable function.

Turing's and Webb's argument by cases runs as follows:

1. The set of states of mind of a person is either finite or infinite.
2. If a person has a finite number of states of mind, then only finitely many may be employed for rule following (since any state of mind a person uses must be a state of mind they have).
3. If a person has an infinite number of states of mind, then only finitely many may be employed for rule following (otherwise the states will be "arbitrarily close and confused").
4. Therefore, by cases, only finitely many states are used by a person in computing.

Clearly, if we reject the disjunction that all sets are either finite or finite, we can reject the first premise and the argument fails. As noted previously, the first premise cannot be justified by appeal to the law of the excluded third. I want to point out that finite is being interpreted as "strictly finite". Let C(m, n) mean "m states are used in computing on input n" by some person. The conclusion that Turing and Webb want to draw, contra Gödel, is

$$\exists n \forall m C(m,n)$$

I argue that this fails, and instead the following is the case:

$$\forall k \exists n \exists m [\diamond C(m,n) \wedge n \succ k]$$

This second sentence doesn't rule out that on a given input, a person only uses a finite number of states in computing. However, it allows that on different inputs, it is possible to use an unbounded number of states—which is implied by having a potential infinity of states.

Recent commentators have placed emphasis on other comments of Turing's to provide support for the finitude of the human mind. Shagrir and Pitowsky argue

> The constraints on [the operations of a human calculation] are virtually self evident. Thus, for example, [the requirement that the number of states of mind which need to be taken into account is finite] should not be understood as the claim that the number of states of mind in general is finite. It merely asserts that the number of states of mind germane to the calculation is finite. As Turing noted, this follows from the fact that we could "avoid introducing the "state of mind" by considering a more physical and definite counterpart of it" (Turing 1936: 139), such as a finite list of written instructions. (Shagrir and Pitowsky 2003: 96)

At first, the authors quoted seem to be sympathetic to Gödel's claim that a human mind is "constantly changing." Gödel's intuition does not deny that, for a given computation on a particular input, only a finite number of states may be used. However, his point is there is no bound on the number of states which may be used for all computations on all inputs. In arguing for the finitude of states of mind, Shagrir and Pitowsky appeal to Turing's comment that we could write down whatever information was needed and run a Universal Turing machine. However, to argue from this to the finitude of relevant states is circular. Only on the assumption that the number of states relevant to all possible calculations on all inputs is bounded and finite can we write out a finite list of equivalent external instructions.

In a moment we will discuss what happens when we deny that the number of states that a computer, human or otherwise, can be in is finite. First, we will cover some contemporary discussion of computability to see what is at stake in doing so.

Supertasks and Hypercomputation

Hamkins (Hamkins 2002) gives a mathematical model of "Infinite Time Turing Machines" and a simple algorithm for computing the halting function on this model. Hamkins' Infinite Time TuringMachines behave as normal Turing Machines do, except for the added definition of a limit state, which defines behaviour for the machine at limit ordinals of computation steps completed. Copeland, in his description of accelerating Turing Machines (Copeland 2002), gives us a model for implementing Infinite Time Turing Machiness. Have the machine complete its first step in half of a second, the second step in a quarter of a second, and so on. Then have the machine behave after a second as the Infinite Time Turing Machines does at the first limit state. If it does not halt, then have it continue for another second, and so on.

A **supertask** is the completion of an infinite number of tasks in a finite amount of time. Clearly, if we could complete supertasks, then we would be able to to hypercompute. For, we could could apply Hamkins' algorithm using Copeland's model and finish in a finite amount of time, thus solving the halting problem.

The completion of a supertask, however, seems to imply paradoxes. James Thomson's article "Tasks and Super-Tasks" (Thomson 1954-5) presents some reasons why. Here is one problem with completing a supertask. Suppose you have a lamp with a switch which turns it from off to on, and on to off. Suppose you can pull the switch in half the time it took you the last time. So, suppose the lamp starts at the off setting. Then, after one minute it is on. After a minute and a half it is off again.

At the end of two minutes, is the lamp on or off?
It can't be on, because it was never turned on without then being turned off.
It can't be off, because whenever it was off, it was then turned on.

Paul Benacerraf (Benacerraf 1962) argues that Thomson here holds a premise which he cannot justify, namely, that the state of the lamp at the end of the supertask is a logical consequence of what happens to it in the moments before the end of the interval. Earman and Norton (Earman and Norton 1996) seem to agree with Benacerraf's general analysis of Thomson's paradoxes, and claim that he has not shown that supertasks are in general impossible. However, they cite another of his paradoxes which, they claim, shows that no computing machine can complete a supertask. In short, imagine a computer which rewrites a single memory address at every computation step. What value does the address have once the computer has completed its supertask?

Notice, however, that Hamkins builds a solution to this worry into his specification of Infinite Time Turing Machines. In fact, the solution that he employs is just the one that Benacerraf, and Earman and Norton might expect based on their analysis of Thomson's lamp paradox. Hamkins simply specifies that at the limit state of his machine, cells which do not have a limit state are arbitrarily assigned the value 1. That is, if a cell changes from 1 to 0, and vice versa infinitely often, then we ensure that the limit state of that cell is logically independent of its state at moments arbitrarily close in time to the limit state. Despite the coherence of hypercomputation through supertasks, it is still questionable whether anything or anyone actually completes supertasks. In the next section I describe a model of hypercomputation independent of supertasks.

Potentially Infinite Computers and Hypercomputation

A potentially infinite computer is just like a regular desktop PC—it has memory, an executive unit, and various input and output devices. However, it adds the following relaxation on existing computers: not only can the memory be increased arbitrarily and without bound, but the executive unit, too, can grow and change without any fixed limit.

For those of us familiar with the Turing Machine model, imagine keeping every part of it the same, except allowing the set of states to be potentially infinite. It turns out that machines so constructed can compute any function over the natural numbers, including the halting function. In rough sketch, we first show that the characteristic function of any language over the natural numbers is computable by Potentially Infinite Computers. Then it follows almost immediately that we can compute any function over the natural numbers. I prove this result formally elsewhere.[58]

What about the next claim, that the model we end up with makes no use of a supertask? Consider a potentially infinite computer at work on some input. If it ever produces an output, then we can be certain that it has in fact used both a finite amount of memory, and that it has used a finite number of states. The potential infinity of the number of states merely tells us that we can't beforehand predict a limit on how many states the potentially infinite computer will use. Arguing for the third claim, that potentially infinite computers are a plausible model of hypercomputation will take a few steps. First, I will argue that they are as plausible a model of real computing objects as the Turing Machine model of computation. Then I will argue that both my model, and Turing's model, both describe actual computing objects, against a famous criticism.

To see that potentially infinite computers are as plausible as regular computers, consider Turing's comments about the implementation of his model in actual objects, in his 1954 *Mind* article.

> Most actual digital computers have only a finite store. There is no theoretical difficulty in the idea of a computer with an unlimited store. Of course, only a finite part of it can have been used at any one time. Likewise, only a finite part of it can have been constructed, but we can imagine more and more being added as required (Turing 1950, 35).

Notice that the adding function over the natural numbers is infinite. We can nevertheless agree that desktop PCs really do compute the adding function by following Turing's comments. For any two numbers we wish to add, if we don't

[58] See Darren Abramson, Ph. D. Thesis.

already have enough memory, then we can provide it with more to complete the computation.

Notice that it will not do to complain that the executive unit, or processor of the computer is "truly a part of the computer" in a way in which the memory is not. To claim so is false. Turing, in the various articles cited, takes great pains in telling us that the memory is part of the machine. And, as I mentioned earlier, limiting the memory seriously limits the functions which can be computed. So, to the extent that we think that actual objects are computationally equivalent to Turing Machines, we ought to think that actual objects can implement potentially infinite computers, and thus hypercompute.

However, we may take the obvious other choice, and deny that any objects really do implement Turing Machines. My first response to this is simply to note that our intuitions about everyday objects are seriously violated under this condition. If we truly believe this, then we must deny that calculators and computers add, subtract, multiply, and divide. Perhaps even more seriously, we must deny that we add, subtract, multiply, and divide. Perhaps the most well-known proponent of such a view is Saul Kripke.

In his discussion of Wittgenstein's private language argument, he writes,

> [If] the dispositionalist attempts to define which function I meant as the function determined by the answer I am disposed to give for arbitrarily large arguments, he ignores the fact that my dispositions extend to only finitely many cases. If he tries to appeal to my responses under idealized conditions that overcome this finiteness, he will succeed only if the idealization includes a specification that I will still respond, under these idealized conditions, according to the infinite table of the function I actually meant. But then the circularity of the procedure is evident. The idealized dispositions are determinate only because it is already settled which function I meant (Kripke 1982: 28).

Kripke notes first that since we only in fact live to a finite amount of time, the total set of our dispositions with respect to adding situations is likewise finite. Second, if we appeal to a possible world in which I live for an infinite amount of time in order to explain what I'm doing in this world, then we must simply stipulate that whatever goes on in that world is the meaning of the function I claim to compute in this world.

Valerie Hardcastle attempts (Hardcastle 1996) to respond to Kripke's challenge of non-circularly explaining how it is that people compute functions like addition and subtraction. To do so, she imagines first that we can specify some core properties of "normal physical functioning" for people who compute subsets of addition and subtraction. Then, she supposes that by specifying our behaviour in a possible world in which physical laws are slightly different, and things don'tage or break down, we can say non-circularly what it is we are doing in this world.

I am sympathetic to the general form of Hardcastle's strategy. It has the strength of denying what Kripke presumes throughout his skeptical attack on non-finite human computation—namely, that facts about the actual world are the only ones that may be appealed to in explaining what people do. However, Hardcastle falls prey to the following dichotomy. Since no one finite world will fix the truth of my computational abilities, we must therefore appeal to some infinitary world in which I live for ever to explain what I do. A more straightforward solution, especially given Turing's comments about the non-problematic implementation of his machines, involves appealing to possible worlds in which I live just a little bit longer, and am presented with a few more problems in arithemetic.

Finally, we should note that Kripke's view implies a strict finitism which—as he recognizes—has various other counterintuitive consequences, including that what when we talk about the natural numbers, we are merely playing language games with no real criteria for truth and falsehood. I do not wish to begin a discussion of the foundations of mathematics at this point. However, it seems to be a least a certain amount of success to have shown that the only position in which one can deny the plausibility of potentially infinite computers, and thus hypercomputation, is a Kripkean finitism. Certainly given this choice, accepting that hypercomputation is possible is a much easier decision than having to accept that people can compute supertasks.

In conclusion, I have shown that we have powerful intuitive motivation for accepting that some sets are of neither finite nor infinite size. Once we see this, certain arguments become suspect, including Turing's argument for the adequacy of his model of computability. We can use these sets to construct models of hypercomputation which are on the one hand more plausible than ones which require the completion of supertasks. Furthermore, the model of computation we end up with is, in principle, at least as significant as Turing's model.

References

Benacerraf, P. 1962. Task, super-tasks, and the modern eleatics. *Journal of Philosophy* LIX: 765–784.

Copeland, B. J. 2002. Accelerating turing machines. *Minds and Machines* 12: 281–301.

Earman, J. and J. D. Norton. 1996. Infinite pains: the trouble with supertasks. In *Benacerraf and his critics,* 231-261. Blackwell.

Ellentuck, E. 1965. The universal properties of dedekind finite cardinals. *The Annals of Mathematics*. 82(2): 225–248.

Hamkins, J. D. 2002. Infinite time turing machines. *Minds and Machines* 12:

521–539.

Hardcastle, V. G. 1996. *How to Build a Theory in Cognitive Science.* Albany, NY: SUNY Press.

Kripke, S. A. 1982. *On Rules and Private Language.* Cambridge, MA: Harvard University Press.

Martin Davis, ed. 1965. *The Undecidable: Basic Papers on Undecidable Propositions, Unsolvable Problems and Computable Functions.* Raven Press.

Moore, A. 2001. *The Infinite.* 2nd ed. Routledge.

Putnam, H. 1991. *Representation and Reality.* MIT Press.

Shagrir, O. and Pitowsky, I. 2003. Physical hypercomputation and the church-turing thesis. *Minds and Machines* 13: 87–101.

Thomson, J. 1954-5. Tasks and super-tasks. *Analysis*, 1–13.

Turing, A. 1936. On computable numbers, with an application to the Entscheidungsproblem. *Proceedings of the London Mathematical Society* 45: 230–265.

Turing, A. M. 1950. Computing machinery and intelligence. *Mind* 236: 433–460.

Wang, H. 1974. *From Mathematics to Philosophy.* New York Humanities Press.

Webb, J. C. 1980. *Mechanism, Mentalism, and Metamathematics.* Kluwer.

Williams, B. 1976. The Makropulos case: reflections on the tedium of immortality. In *Problems of the Self: Philosophical Papers 1956-1972,* 82–100. Cambridge University Press.

CHAPTER THIRTEEN

PHILOSOPHICAL FOUNDATIONS FOR A UNIFIED ENTERPRISE MODELLING LANGUAGE

GERALD R. KHOURY
INSTITUTE FOR INFORMATION AND COMMUNICATION TECHNOLOGIES, UNIVERSITY OF TECHNOLOGY SYDNEY

AND SIMEON J. SIMOFF
INSTITUTE FOR INFORMATION AND COMMUNICATION TECHNOLOGIES, UNIVERSITY OF TECHNOLOGY SYDNEY

Introduction

While enterprise architecture (EA) is still in its infancy as a research area, this has not prevented it from becoming an important and firmly established discipline within the IT industry. In fact, EA's are growing in importance as tools for managing change within today's highly dynamic, demand driven and highly competitive business environments. As the rate of technological change increases and information environments become more complex, more sophisticated methods are needed to manage these environments effectively. EA's help manage this change and overcome the problems of building isolated IT solutions that fail to support an enterprise's vision, goals and objectives. "Understanding and visualising complex businesses enables you to identify and address areas that might be constraining business performance. Enterprise modelling helps you focus on those areas you can change, how these areas are currently functioning, how they might be optimised, and how any changes might impact other areas."

However, there are barriers to achieving EA's full potential. Traditional EA models are often unwieldy and difficult to navigate and explore. The problem is exasperated by the fact that the stakeholders to which these architectures are directed have "varied backgrounds, and technical and non-technical skill sets and interests." Typically, these stakeholders include the CIO, a wide range of business users, system users, IT developers, systems analysts and systems architects. In fact, Bemelman and Dennis's investigation of architecture from a users' point of view (in), concludes that "the inherent levels of complexity and detail in most of the current architectures will become a major impediment to acceptance of these architectures in the industry."

The Difficulty with Enterprise Modelling

The ability to create *effective* EA's rests on the ability to create "simple models that are easily communicated…." Unfortunately, current enterprise models are complex, difficult to produce and hard to understand because they can only be produced using variety of unrelated modelling languages. While specialised languages have been developed to model specific domains such as application, infrastructure or network architectures, none of these languages support the creation of high-level, conceptual systems models that extend across the enterprise. "The range of phenomena addressed by enterprise modelling stretches multiple disciplines, and accordingly many modelling languages and practices are used. There is no single person/profession who would be able to guarantee the consistency of all models produced…." Yet, successful systems implementations depend on the availability of integrated models that are consistent and non-contradictory.

In fact, there are currently no standards to guide the development of EA models: "Today, no single existing modelling language by itself is capable of modelling all necessary aspects of an enterprise." This leads to several problems. Firstly, enterprise architects are required to have expertise in multiple modelling languages in order to create enterprise models. Secondly, the lack of consistent semantics and a weak ontological foundation makes it difficult to create coherent and consistent enterprise models. Lastly, the resulting models are beyond the comprehension of those who are not modelling experts and yet are interested in EA. This includes technology and business leaders, perhaps the most important users of EA's.

Ultimately, the use of multiple EA languages that cannot be effectively integrated, prevents the achievement of the prime goal of an EA: the development of an *integrated* view of the enterprise. In particular, EA's that are built using component-based frameworks (such as the popular Zachman framework) are fundamentally flawed because they model the enterprise as a set of independent structures with discrete boundaries. This leads to an EA model that provides a fragmented view of the enterprise with poor explanatory power and little flexibility for future planning.

There *have* been serious attempts to develop a single language that can be used to describe the entire EA domain. Unfortunately, these attempts have suffered from significant drawbacks that have prevented them from achieving industry acceptance. One of the most notable proposals is the "conceptual graph" notation. Zachman and Sowa proposed the use of conceptual graphs, a form of symbolic logic, to provide a natural language that can be used as a unifying ontology for all the models required by the Zachman framework.

However, Zachman and Sowa first suggested the application of conceptual graphs to enterprise modelling more than a decade ago. Yet, there is scant evidence of the application of conceptual graphs to real world EA problems. The primary reason for this is probably that conceptual graphs are too *complex* for general use as an effective modelling tool. This is evidenced by the work of Polovina, who investigated the suitability of conceptual graphs as a way of modelling strategic management problems within an accounting domain. Polovina found that "However, despite their strong prima facie attractiveness ... the inherent complexity of conceptual graphs fundamentally undermined them as a viable tool, other than for very trivial problems well below the level needed to be viable for strategic management accountancy." Although Polovina was working in a specialised domain, there are many similarities between the modelling he was testing and the modelling of EA's. In addition, the senior practicing accountants who were his experimental subjects have similar analytical competence to enterprise architects.

Other attempts to achieve an integrated enterprise modelling language include the Hatley/Pirbhai (H/P), Quantitative QFD (Q^2FD) and WfMC Meta-model approaches. However, all of these methods are highly restricted in terms of domain. As a result, they are unable to deliver effectively the benefits that would be forthcoming from a method that integrated well over the entire EA domain. Also, "even in these domains the models are not in very wide use ...."

It should also be noted that contemporary EA approaches have been developed informally. This suggests that a scientifically grounded, theoretically sound approach to EA development may offer improvements over contemporary EA methods and frameworks.

In this paper, a new theory is expounded that can be used to develop an integrated approach to EA modelling. This theory makes use of an understanding of metaphor and their relation to system model hierarchies. The following section provides an understanding of the place of metaphor within type hierarchies. We then look at the connection between models and metaphors and this provides a link between type hierarchies, metaphor and model hierarchies.

Dynamic Type Hierarchies

The dynamic type hierarchy (DTH) theory "involves a theory of metaphor that incorporates Sowa's conceptual graphs, dynamic type hierarchies, and Max Black's interaction approach." Numerous theories have been developed explaining how metaphors are used and why they work. These theories include emotive theories, the substitution approach, the comparison theory, metaphor as analogy, the controversion theory and the anomaly theory. Way surveys and analyses these approaches and illustrates the shortcomings and contradictions that beset them. Way finds that the interaction view, while not perfect, is the most promising contemporary theory of metaphor and that "much of the experimental data of metaphor is either compatible with or actually supports aspects of the interaction view."

Max Black originated the interaction view. Way summarises the main points of the interaction view of metaphor as follows:

> metaphor involves entire systems of assumptions and "commonplaces" which are associated with the terms involved;
> that the metaphorical process works like a filter, with the associated ideas of the secondary subject (vehicle) hiding, highlighting and organizing aspects of the primary subject;
> understanding metaphor often involves a shift in meaning;
> metaphor cannot be reduced to any literal statements of comparison, and;
> metaphor can actually create similarity between previously dissimilar ideas.

Way acknowledges that there are criticisms of the interaction view, however, she believes that despite these "the interaction view has been the most widely accepted and influential view of metaphor." Subsequently, Way's new theory, the DTH theory, is developed as a variation and improvement on the interaction view theory. The DTH theory is thus purported to explain the results of empirical tests where all other theories fail.

The DTH theory uses standard concepts from graph theory: "A type hierarchy is a network of concepts which are organized according to levels of generality. … So the links connecting the supertypes and subtypes of the semantic network represent going from an instance of a supertype to a more specific instance of that supertype. … an instance of a subtype entails that it is also an instance of the corresponding supertype." The DTH is a generalisation hierarchy where the entities are categorised based on the common attributes. In a generalisation hierarchy, the higher-level class (supertype) share the common attributes of the lower level class (subtype), while subtypes inherit all the properties of its supertype.

"The DTH theory holds that the similarity found between the tenor and vehicle of a metaphor[59] is not an intersection of their properties; rather, it is generated by finding a common and more abstract *supertype* that the two share." The supertype is a generalisation of the attributes of the connected subtypes. "In metaphor, what is common between the vehicle, and tenor is not an intersection of a list of features at the level of the tenor and vehicle, but a supertype, which is higher up on the semantic hierarchy and under which aspects of both the vehicle and tenor domains fall. Furthermore, which supertypes are chosen, assuming that there are several in common, is a function of the context and the direction of the attribution of the metaphor, that is, the metaphor is attributing features from the vehicle domain by abstracting them to a common supertype and then using that supertype to pick out the corresponding features of the tenor."

According to the DTH theory, metaphor operates through a higher-level supertype that connects the tenor and vehicle types. If this supertype is not found to already exist in the type hierarchy, it is dynamically created, just as in some cases "metaphor creates the similarity" rather than describing some existing similarity. Two separate clusters of schemata thus become associated through a supertype and this allows a migration of concepts from the vehicle to the tenor. This exemplifies what Black terms "systems of associated commonplaces:" the complex and far reaching web of associations and meaning that imbue every metaphor.

Way provides as an example the metaphor "the car is thirsty". This metaphor "involves a violation of a constraint, in that thirsty is an attribute of an animal, not a vehicle. The new supertype, *Mobile-entities that require liquid*, can have both tenor and vehicle fall under it without violating any semantic constraints." Note that the set of properties of the supertype will be smaller than those of its subtypes; in this way the supertypes are able to act as filters on lower level concepts. ... The supertypes VEHICLE picks out one aspect of a car, while another supertype, say STATUS-SYMBOL, would pick out or filter different aspects.

Models and Metaphors

Lakoff and Johnson define metaphor as: "... understanding and experiencing one kind of thing in terms of another." Their ontology parallels the description of metaphor used widely within contemporary studies of cognitive linguistics, where the terms source and target refer to the conceptual spaces connected by the metaphor. The structure of the source domain is projected onto the target domain in a way that is consistent with the inherent target domain structure.

---

[59] Here, the terms tenor and vehicle can be reasonably substituted for target and source respectively.

Elsewhere in the research literature, the target is variously referred to as the primary system, topic or tenor, and the source is often called the secondary system or the vehicle.

Contemporary research has shown us that metaphor, far from being just a figure of speech, is central to everyday communication and learning. In fact, it has been argued that only by relating a new concept (the "target") to a well understood, everyday object (the "source"), can one develop new knowledge and understanding and that without this process, it is impossible for humans to carry out abstract thinking.

As our understanding of metaphor has developed, the links between models and metaphor have become more apparent. This is most evident in the case of linguistic models. Black was one of the first to make the case for the connection between models and metaphor in language and philosophy, and in later years, Black becomes further impressed "by the tight connections between the notions of models and metaphors. ... Every metaphor is the tip of a submerged model." More recently however, these early notions on the connection between models and metaphors have been extended. Lakoff, for example, demonstrates that mathematical models are also metaphors. According to Roussev and Rousseva, the mechanics of modelling are the same as metaphor, and modelling evokes the same cognitive processes as metaphor. Indeed, "Models are always grounded in a kernel metaphor, or metaphors, and hence stimulate many of the same cognitive processes."

Interestingly, the definition of a model—"A model is a representation of something else"—is remarkably similar to the definition of a metaphor as an "image that represents one thing as something else in order to explain it better." In fact, an *empirical* argument can be made for the assertion that metaphors are models. According to Herbert Stachowiak (as referenced in an artefact must satisfy the following three criteria in order for it to be identified as a model:

> Mapping criterion: there is an original object or phenomenon that is mapped to the model.
> Reduction criterion: not all the properties of the original are mapped on to the model, but the model is somehow reduced. On the other hand, the model must mirror at least some properties of the original.[60]
> Pragmatic criterion: the model can replace the original for some purpose, i.e. the model is useful.

---

[60] It must be noted, that this is contrary to some popular notions of models. For instance, Ghyczy (2003) suggests that models "exhibit a one-to-one correspondance" with their source and that you can "transfer everything you know about the source domain into the target domain" if you have a good model. Clearly, this is false: models are only useful because they are *abstractions* of the source domain, and by implication, this means that everything must *not* be transferred from the source to target domains.

It will be recalled that a metaphor connects a source domain to a target domain. This satisfies the first criterion. When metaphor is used, only certain characteristics of the source are mapped to the target, and this depends upon the context in which the metaphor is being used. Thus, the second criterion is satisfied. Finally, we use metaphor in order to explain something better, especially when that something is new and novel. Therefore, metaphor satisfies all three of Stachowiak's criteria defining a model, and metaphor is a type of model. More specifically, metaphor is a type of conceptual model.

Furthermore, it can be observed that the labels given to enterprise systems are actually metaphors (and by induction, models). Take as an example, an enterprise system labelled "Customer Relationship Management System." If this were simply a label, and not a metaphor, then there would be no significant effect in changing this label to "Customer Exploitation Management System" as long as there were no changes made to the actual systems to which this model refers. But in fact, this label change fundamentally changes the perceptions that will be generated around this system. A completely different set of "associated commonplaces" has been set up, because these labels are not simply literal: they are metaphors which give rise to supertypes which encapsulates much of the understanding of what the system is and how it operates and so an "entire web of associations and implications" has been altered.

Therefore, the *labels* that are assigned to enterprise systems, can themselves, be used as models if they are terms that have associated meanings outside of the system context.

In fact, the metaphorical approach to organisational modelling is founded on the observation that the reality of organisational structures are not concrete, but abstract, indefinite, and perhaps even undefinable, entities. There is no objective reality of an organisation. The significant structures within an organisation are actually social constructs, and as such, they are imbued with cultural and social meaning, presenting differently to every person according to their immediate concerns. In modelling organisations we are not simply developing abstractions of an objective and concrete structure. Rather, we are saying "this is how we *wish* the organisation to be understood." For example, "we want our enterprise to be seen as developing relationships with customers, not exploiting them." As Black says, the metaphor actually helps constitute aspects of reality and we create meaning through the development of metaphors for these structures.

Model Hierarchies

There are two conditions for the development of an effective unified model. Firstly, the unified model must be capable of representing the semantics of the various sub-domains that are being modelled, albeit, at a higher level of abstraction. Secondly, the unified model must present a strong ontology so that the model can be interpreted unambiguously. A *single*, well-defined and formalised modelling language will provide a strong ontology. Unfortunately, no current modelling languages can cover the *semantic* breadth needed to describe the different types of enterprise systems, without inordinate complexity. In this section we deal with the problem of developing a semantically unified enterprise model. In the following section, this model is used to develop an ontology that can be applied to formally describe any sub-domain.

It has already been shown that metaphor can be viewed as part of a dynamic type hierarchy. For the purposes of this argument, the types in this hierarchy will be restricted to the domain of enterprise models. Note that, since it has been shown that metaphors are a special type of model, this restricted domain still *includes* metaphor.

Within any enterprise, there is a set of models representing various levels of abstraction. At the highest level is the model of the enterprise itself. This may, or may not be formalised. At lower levels of abstraction are more detailed models such as system and function models. It can be observed that larger scale structures tend to be modelled using more general, "conceptual" models and metaphors while smaller scale structures tend to be modelled using more specific, "concrete" models and metaphors. Subsequently, enterprise system models can be formed into a generalisation hierarchy, where the higher-level

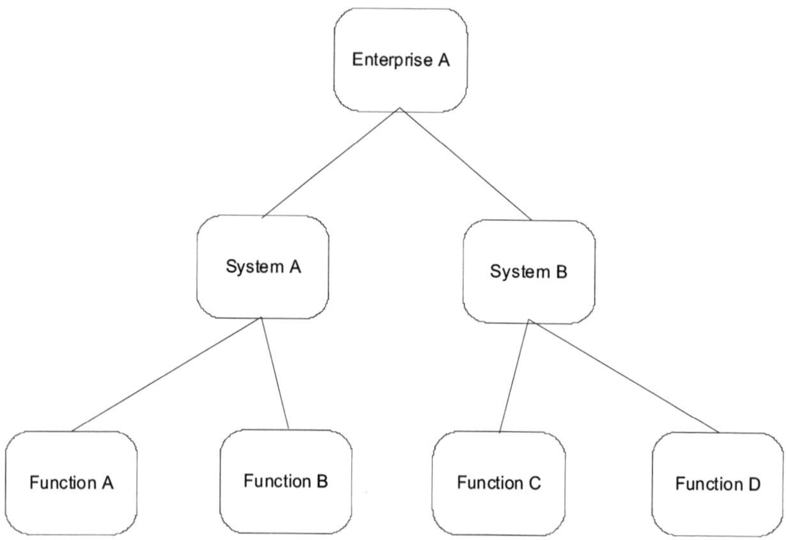

Figure 1. Enterprise System Hieararchy with Enterprise as Global Supertype

classes (supertypes) share the common attributes of the lower level classes (subtypes), while subtypes inherit all the properties of its supertypes.

Figure 1 shows an enterprise system hierarchy where “Enterprise A” is the universal supertype. According to Way, “metaphor cannot take place by comparing the properties of independent systems on the same level,” i.e. at the same level of abstraction. “In order to properly make the comparison, we must search for higher order concepts. Comparison of properties on one level only make sense as a comparison with respect to a common, more abstract, higher level property.” Therefore, it is possible to create a metaphor between “Enterprise A” and another type at the same level of abstraction as Enterprise A that could be used to describe features of the enterprise and each of its component structures (“System A”, “System B”, “Function A”, “Function B” etc) by generating a higher level type. However, as every enterprise is unique, this metaphor may not be relevant to other enterprises or their component structures. For example, we might say, “Enterprise A *is* transportation”. If “Enterprise A” was, say, an airline company, this metaphor could provide a useful ontology and a desirable set of associated meanings and implications. However, if “Enterprise A” were a finance company, then this metaphor would not be useful or desirable.

On the other hand, if we created a metaphor that compared a *supertype* of an enterprise to another type on the same level as that supertype, then the metaphor *would* be relevant to all enterprises. In fact, it would provide a unified concept that could be applied to any of its sub-models. This is illustrated in Figure 2.

The task remains, to identify, or invent, a supertype of all enterprise systems models.

There are any number of abstract concepts that could be used as a metaphor source for enterprise systems. These include the description of organisations as machines, organisms, brains, cultures, political systems and learning organisations. Note however, that some of these metaphors use source objects that are at the same, or lower levels of abstraction than the enterprise type. Therefore, they are not suitable as a unifying metaphor for all enterprises. Take as an example, the description of an organisation as a machine. This may describe certain aspects of an organisation, however it is a highly concrete metaphor and is too confining to describe all aspects of all systems. For instance, it cannot describe the ability of an organisation to learn, regenerate or develop purpose. The metaphor of “the learning organisation” is however more conceptual and flexible, explaining perhaps it greater popularity as an enterprise metaphor. Another highly conceptual metaphor is “the enterprise as a societal structure". In the following section, this metaphor is adopted and developed as a basis for a unified systems language.

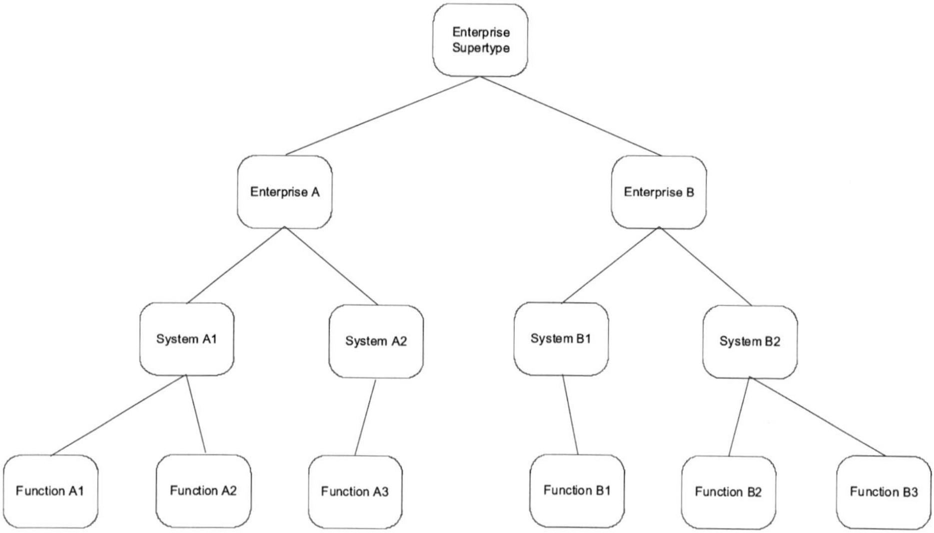

Figure 2.Enterprise System Hieararchy with New Enterprise Supertype

A Societal Language for Unified Systems Modelling

Perhaps one of the conceptual structures most familiar to humans is that of human society. Society is a larger scale structure than an enterprise (societies contain enterprises, but enterprises aren't normally viewed as containing societies), and accordingly, it can be used as a supertype of all other types in the enterprise system type hierarchy. That is, a society-sourced metaphor will be at a higher level of abstraction than any of the models that describe an enterprise or its components. Therefore, the concept of society has the semantic breadth to describe all enterprise systems.

Is a society-sourced metaphor a good choice for modelling enterprise systems? It is clear that all enterprises are social phenomena and that computer systems are socially embedded phenomena. In lieu of a social context, computers can have no meaning or value. It is also widely noted that many systems failures occur, not because of technical issues, but because the relevant socio-political concerns were not evaluated and addressed. A society-sourced metaphor is likely to be stronger than many other metaphors for modelling these aspects of the enterprise. Thus the "pragmatics" of this metaphor[61] appear to be strong, especially as any society based concepts and nomenclature are likely to be very familiar to all users of the enterprise architecture. While there *are* other metaphors that would satisfy the semantic requirements of a unifying supertype, the concept of society is perhaps a more promising metaphor for this problem, since it has a particular relevance and efficacy for enterprise systems modelling.

Formalisation and codification of the enterprise model supertype will provide a foundation for a modelling language that can be used to describe component structures of any enterprise, but to do this we need to understand what a model of society looks like. Fortunately, one such model has been developed by Giddens. According to Giddens' Theory of Structuration, there is interdependency between humans (actors) and societal structures (resources and rules) that is manifest through specific actions.

The notion of an actor is extended here to include, not only individuals, but also any agent that can exert power in order to produce an effect. To this end, the terms actor and agent are used interchangeably. Resources are "structured properties of social systems, drawn upon and reproduced by knowledgeable agents in the course of interaction." Resources are of two types. Allocative resources are material resources involved in the generation of power and derive from human dominion over nature. Authoritative resources are non-material and derive from the capability of harnessing the activities of human beings. Rules refer to the sanctioned modes of conduct, and an action is an activity that is performed. "Structures, as 'rules and resources,' do not do anything, but they have their effect through being known and used by actors."

This provides an ontology consisting of actors, resources, rules and actions that can be used as the basis for developing a formal modelling language. These four structures might be used as nodes, for instance, in a graphical language that connects these nodes using arcs according to well-defined rules. Furthermore, the nodes themselves could be used as the minimal common supertypes of four different type hierarchies, where the subtypes within each of these hierarchies inherit the attributes of its parent. This would provide the language with a high level of semantic power, despite the fact that only four normative types exist.

Conclusion

"Enterprise models are useful only if they are used. They will be accepted by users as a tool if they are simple to understand, easy to use, computer-supported, and if they provide a realistic image of the reality. This explains the

[61] A measure for the degree to which models expressed using this metaphor will succeed in being interpreted by the audience as intended by the creator.

failure of many approaches proposed in the past."

Without a *unified* enterprise modelling language, there is no way to develop coherent and consistent enterprise wide models of IT systems. Thus, the development of a *unified* EA modelling language is a highly coveted goal. Its achievement would overcome some serious shortcomings with current EA approaches and improve the efficiency and effectiveness of enterprise architects and systems planners. Unfortunately, the development of such a language has remained elusive.

A philosophy-based approach to the problem of unified EA modelling has the potential to solve a problem that has proven to be intractable using more traditional and direct approaches. By applying an understanding of type hierarchies to the selection of system metaphors, it is possible to identify a metaphor that provides a scope that is wide enough to cover all of the functional elements of the enterprise's component structures. After all, it is better to have a single metaphor that covers the entire domain, for "when discourse becomes full of conflicting metaphors, it may be difficult for the uninitiated to keep their bearings."

This metaphor can then be used to develop a unified language that can model all of the enterprise's systems. Since a single metaphor is used to replace a myriad of disparate, and possibly, contradictory metaphors, the resulting models have the following qualities:

Greater explanatory power (a consistent metaphor/language is used to model the entire organisation which allows the user to develop an effective mental map of the organisation).

Greater flexibility for the management of change and strategic planning (current disjunctions between systems, business units etc. are not "hard wired" into the model).

Avoiding loss of information that might occur in translating from one architectural view to another and ensure cross-view consistency.

Explicit capture and representation of business needs within the EA models

Reducing cognitive load placed on a user that needs to understand a complex set of architectural views

Research is continuing in this area and includes the development of a tool to support the application of these languages, as well as the evaluation of unified languages that are generated using the theory described in this paper for "real-world" enterprise architecture problems.

References

Allen, R. B. 1997. In *Handbook of Human-Computer Interaction*, eds. M. Helander, T. K. Landauer, and P. Prabhu. Elsevier Science.

Armour, F. J., S. H. Kaisler, J. Getter, and D. Pippin. 2003. A UML-driven enterprise architecture case study *System Sciences, 2003. Proceedings of the 36th Annual Hawaii International Conference on* Hawaii.

Bernus, P., L. Nemes, and R. Morris. 1996. In *Modelling and Methodologies for Enterprise Integration*, eds. P. Bernus and L. Nemes. London: Chapman and Hall.

Biemans, F. P. M., M. M. Lankhorst, W. B. Teeuw, and R. G. v. d. Wetering. 2001. Dealing with the Complexity of Business Systems Architecting *Systems Engineering* 4: 118-133.

Black, M. 1962. *Models and Metaphors: Studies in Language and Philosophy.* London: Cornell University Press.

—. 1979. In *Metaphor and Thought*, ed. A. Ortony, 19-43. London: Cornell University Press.

Fraser, J. and Tate, A. 1995 The Enterprise Tools Set - An Open Enterprise Architecture *Proceedings of Workshop on Intelligent Manufacturing Systems (IMS), International Joint Conference on Artificial Intelligence (IJCAI-95).*

Ghyczy, T. v. 2003. The fruitful flaws of strategy metaphors. *Harvard Business Review*: 86-94.

Giddens, A. 1984. *The Constitution of Society: Outline of the Theory of Structuration.* Cambridge: Polity Press.

Johnson, G. J. 1994. Of metaphor and the difficulty of computer discourse. *Communications of the ACM* 37: 97-102.

Lakoff, G. 1993. In *Metaphor and Thought*, ed. A. Ortony, 202-251. Cambridge University Press.

Lakoff, G. and M. Johnson. 1980. *Metaphors we live by.* Chicago, IL: University of Chicago Press.

Lakoff, G. and R. E. Nunez. 2000. *Where Mathematics Comes From.* New York: Basic Books.

Ludewig, J. 2003. Models in software engineering - an introduction *Software and Systems Modelling,* 2: 5-14.

Maier, M. W. and E. Rechtin. 2000. *The Art of Systems Architecting.* Boca Raton: CRC Press.

Morgan, G. 1996. *Images of Organization.* SAGE Publications.

Noran, O. 2003. An analysis of the Zachman framework for enterprise architecture from the GERAM perspective *Annual Reviews in Control* 27: 163-183.

Parker, J. 2000. *Structuration.* Buckingham: Open University Press.

Polovina, S. 1993. Loughborough University of Technology.

Rostad, C. C. 2000. In *Enterprise Modeling: Improving Global Industrial Competitiveness*, eds. A. Rostadas and B. Andersen, 119-136. Boston: Kluwer Academic Publishers.

Roussev, B. and Y. Rousseva. 2004. Software Development: Informing Sciences Perspective. *Issues in Informing Science and Information Technology,* 1: 237-245.

Solberg, C. 2000. In *Enterprise Modeling: Improving Global Industrial Competitiveness*, eds. A. Rostadas and B. Andersen, 183-199. Boston: Kluwer Academic Publishers.

Szegheo, O. 2000. In *Enterprise Modeling: Improving Global Industrial Competitiveness,* eds. A. Rostadas and B. Andersen, 21-32. Boston: Kluwer Academic Publishers.

Walsham, G. and Han, C.-K. 1991. Structuration Theory and Information Systems Research *Journal of Applied Systems Analysis* 17: 77-85.

Way, E. C. 1991. *Knowledge Representation and Metaphor.* Dordrecht: Kluwer Academic Publishers.

Winograd, T. and F. Flores. 1987. *Understanding Computers and Cognition: A New Foundation for Design.* Reading, MA: Addison-Wesley.

Zachman, J. A. and J. Sowa. 1992. Extending and formalizing the framework for information systems architecture. *IBM Systems Journal* 31: 590-616.

CHAPTER FOURTEEN

THE HEARING DIMENSION OF 3D VIRTUAL WORLDS: UNEXPLOITED OPPORTUNITIES

LUDMIL DURIDANOV
FSU JENA

AND SIMEON J. SIMOFF
INSTITUTE FOR INFORMATION AND COMMUNICATION TECHNOLOGIES, UNIVERSITY OF TECHNOLOGY SYDNEY

Introduction: Virtual Worlds as Technology and "Living" Environment

Technologically, virtual worlds have started to evolve as a new medium since the beginning of the last century. The US Patent *1183492* for a head-based periscope, awarded to Albert B. Pratt in 1916, is considered to be the first head-mounted display (Sherman and Craig 2003, 24). The seminal works of Ivan Sutherland brought a number of significant developments in virtual reality technology, including the concept of interactive computer graphics (Sutherland 1963) and the concept of ultimate display (a display in which a person can interact with the objects in a world that does not necessarily follows the laws of the physical world) (Sutherland 1965). The emphasis in the design of virtual worlds is on the interaction between people and constructed video representations of the real world. For example, Krueger's *video place* (1982) utilizes video cameras and other input devices to construct a virtual world controlled by the *untethered* human motions. These virtual reality technologies are computationally intensive and expensive applications, used mostly in specific domains, like training systems for air pilot, drivers and other simulations of physical world environment.

The development of distributed digital media, linked through the World Wide Web (WWW) technology, created a radically new opportunity for integration, access communication and sharing of information resources. The combination of virtual worlds concept with distributed Web media and client-server computing led to a *paradigm shift*, namely, to the development of visual 2D and 3D virtual spaces in the 80-s and 90s of the XX century.[62] These spaces have been developed initially as social spaces ("places") for virtual communities. A typical example of such 2D (Habitat) (Rheingold 1991) and 3D (the Active Worlds Alpha Universe)[63] spaces are shown in Figure 3a and 1b, respectively. The *domination* of *"visualisation requirements"* shaped the interfaces to the distributed virtual worlds on the Internet.

Figure 3b illustrates an example of typical interface to the virtual world space. The access includes a 3D area where all the action in the virtual space happens. The 3D area is populated by different objects that constitute the design of the virtual world. The 3D designs are referred also as *virtual architecture.* There is a chat area, where the synchronous text-based communication takes place. The avatars embedded in the virtual world enrich communication through their individual sets of gestures, which vary from avatar to avatar.

1a. 2D virtual world

[62] This was later referred as a *visual revolution* by J. Lanier, one of the fathers of this development (Zaleski 1997 : 134-157, 180-196).
[63] http://www.activeworlds.com

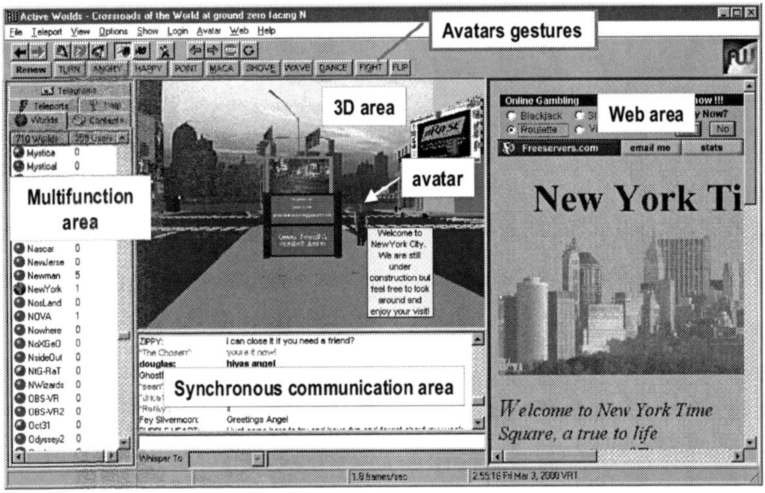

1b. 3D virtual world (Active Worlds)

Figure 3. Examples of the interfaces to 2D and 3D distributed virtual worlds operating over the Internet

The essential lesson learned from the experiences with Habitat is that cyberspace is defined more by the interactions among the actors within it (Qvortrup 2001) than by the technology with which it is implemented. In other words, virtual worlds (in broad sense) to some extent are independent of their medium (the virtual worlds as technology). A virtual world is a real representation of a world that may have or may have not a physical counterpart. It can be thought as a simulation of a *domain* manifested in some medium where the simulation is driven by *rules of behaviour—programmed* (in computer-mediated ones) or *imagined* (in mental worlds)—simple or complex. These rules are directly related to the design of the virtual world. In present these rules heavily affect the forms, colours and dimensions of the visual space. The domain includes the space, avatars, objects and rules of the virtual world. To some extent domain is approximated by the metaphor used to describe the relations of the elements that constitute the domain.

Figure 4 illustrates the designs of virtual offices produced by different set of rules. The design of the office in

Figure 4a follows the rules that simulate "traditional" office setup, including furniture and even a simulation of a desk lamp. The mapping defined by the *metaphoric analogies* of the physical office design has the advantage of the familiarity of the relationships between the objects in the virtual world (e.g. working documents are expected to be found on the desk). The design of the office in

Figure 4b follows the *aesthetic rules* derived from Mondrian's neoplasticism.[64] The key advantage of this design is the novelty and the style that appeals towards *our perception of beauty* and drags the user on to "stay inside" for a while. Such style can cause "*positive vibes*" that can setup the comfortable "*communication wave*" of the people inhabiting such space.

2a. "Traditional" virtual architecture design

2b. Virtual architecture design influenced by Mondrian style (Maher 1999)

---

[64] Piet Mondrian (1872-1944) was a Dutch painter known for his rigidly geometric style, which he labelled as Neoplasticism. The style is characterised by the use of straight lines in horizontal and vertical directions, forming patterns of rectangles, and filling the rectangles with pure primary colours. Mondrian regarded neoplasticism as an attempt to unify the arts and give man an orderly environment of beauty.

Figure 4. Different rules define different spatial (and visual) designs

When experiencing a space with other human participants it is important to *be able to sense* their *presence* in the virtual world, i.e. to be aware where they are positioned, which way they are looking and what they are saying. Avatars[65] can *visualise* humans in a virtual ambiance. They vary from simple images to save on computational expense, through 3D simplified geometries, ending with photo-realistic representations (including a full 3D body scan of a person) as illustrated in Figure 5. The photo-realistic avatar at the right-hand side in Figure 5 is shown within an avatar design studio, which allows manipulating and designing the geometry, look-and-feel, gestures and other animated behaviours in terms of avatar character.

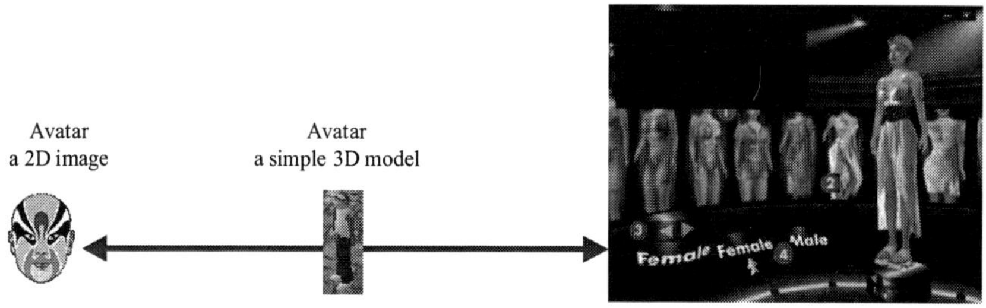

Figure 5. Variety of avatars in distributed virtual worlds

The sound as a medium element in virtual worlds

The design of virtual worlds has been substantially biased towards the "visual" component,[66] reflected in the explosive growth of information visualisation techniques (Chen 1999). The stronger accent has been placed on *"what could be seen."* In the embedded challenge *audio* vs. *visual dimensions* this led to some underestimation of *"what could be heard."* The possible damage by an unbalanced interaction and inhabitancy of virtual reality is caused by the overload of human visual perception system, which may lead to a discomfort and unpleasant perceptions in visually intensive virtual worlds, caused by the not necessarily perception of the virtual world as a seamless extension of the physical world in which the person lives. Consequently, a virtual world inhabitant is confronted with the dilemma to give up this „adoptive identity" and to rush to another one till he returns to *his own* reality. J. Lanier, one of the founders of the *visualized virtual reality* (J. Zaleski 1997, 135)

[65] The original meaning of *avatar* is the earthly embodiment of a deity.
[66] J. Lanier is calling it a "revelation" as well (Zaleski 1997: 137).

explains this phenomenon with the *propelling source energy* of the web, an "*erotic quality*" that turns you on and on to surf away, to remove the next veil, driving you in the state of not being able to stay longer than a couple of minutes at the same site.

Our point of interest is to turn our view towards the sound dimensions as fundamental for the "insider feeling" of virtual reality. The appropriate sound vibrations can greatly enhance person's ability to become mentally immersed in the virtual world. From *ambient sounds* that give cues to the mood of the virtual world setting to sounds associated with particular avatars, *aural representation* is a key to positioning participant in a comfortable zone and offering enjoyable experience.

Further we consider some of the current issues at the sound level. The groundwork for using computer-generated and computer-controlled sound in virtual worlds was laid in the early 80s of the past century, in the area of *sonification* (the use of sound to represent large datasets) (Bly 1982). Sonification could be viewed as the equivalent of information visualisation in the area of sound—the presentation of information in an abstract sound form. Sonification maps a number of parameters of a dataset to specific parameters of sound, so that the human sound system can detect patterns in the data based on the resulting sounds. For example, particular sound can vary based on the changing mood of a person behind an avatar.

At present, sounds in virtual worlds are grouped in (i) *ambient sounds*; (ii) *marker sounds* that mark an event; (iii) *index sounds*—sounds that continually provide information about the state of something, and; (iv) sounds that *augment* or *substitute* for perceptions made by the other senses.

Ambient (or *background*) *sounds* are the sounds generally used to set the mood of an experience. The technique is well understood and applied by film and play directors. The ambient sound technique is not interactive. For example, a common sample loop sound used at the entry of virtual worlds is the sound of birds in a forest or of flowing water. The sound aims at calming and relaxing the people who enter the virtual world. These sounds usually do not change with respect to the user interaction with the virtual world, a part from disappearing when the user gets at a particular distance from the entry to the world. The 3D sound allocation and modelling for virtual reality applications has been initiated in the 90s by Scott Fisher and Elizabeth Wenzel at NASA. Nowadays virtual worlds use *spatialised sound models*. These models emulate the fading of the sound in the atmospheric medium. The idea is illustrated in

Figure 6. Let avatar *P6* be the source of the sound. According to the distance attenuation model the sound should be fading in radial directions from that person, and at a certain distance will become below the threshold for being heard. In our example, avatars *P1*, *P2* and *P3* will perceive the sound, when *P4* and *P5* will not.

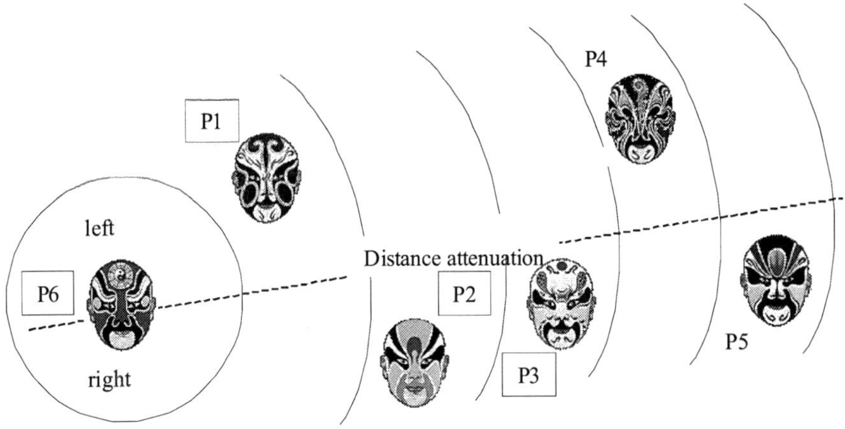

Figure 6. Distance attenuation model for sounds in virtual worlds

Marker sounds are the sounds that indicate the occurrence of a particular event in the virtual world. They denote discrete, fleeting events. The mood-setting nature of the *ambient sounds* can serve as a marker sound. For example, a change in the rhythm of a drum beat from a disco kick to rhythm and blues may indicate a change in the negotiation phase between participants in a deal struck in an electronic markets virtual world. *Marker sounds* may be also a separate set of sounds, including world event indicators, interface event indicators, sonification events and sensory substitution events. For example, a world event could be the sound of a door closing (when an avatar is bumped out of an area), a sensory substitution sound could be a „cracking sound" when two avatars collide.

Index sounds reflect the changing value of a continuous entity, mapping a continuous value (e.g. velocity of movement) to a sonic parameter (e.g., pitch). Another example could be the use of a melody to indicate the comfortable distance between avatars. The changes in orchestration, timbre, and complexity of the synthesized sound reflect how close the avatars are to an optimal (according to some criteria) arrangement.

Although the above discussed sound support in virtual worlds covers variety of functionalities in a virtual world and recognises the significance of aural representation, its potential for an impact in a consistent and deeper holistic way on inhabitants of virtual worlds is still underutilised. In this sense we use the term "unbalanced communication" to stress the dominance of visual communication. The unbalanced communication with virtual reality, based on

WWW is part of a wider framework of the *universal restructuring* of everyday life in a "stream of nomadic movements" which are about to globalise our civilisation (Deleuze/Guattari 1980, 434-537). In computer-mediated virtual worlds the effect of "diving" into the world begins with physical immersion—synthetic stimulus of the body senses via the use of technology, followed by a mental immersion—having the "sense of presence" within the virtual world. Despite the importance of the visual feedback and experience in the world, the harmony between the visual experience and the aural feedback is an essential ingredient of virtual worlds—the key to *opening* the "self" in a *virtual world.*

Opening the "Self" in a Virtual World

Now let's go back to our dilemma:

A. how to restore the balance of sensations in the process of one's opening to virtual realty; and on the other side,
B. how to create a binding field, where the individual will keep the illusion as an insider of virtual reality, overwhelming the "erotic power" to jump to the next site.

The main failure when not taking in account the harmony, beauty and the affect of the "*musical vibrations*" of a *natural* world has been discussed and stressed by physicists (J. Kepler 1619; A. Einstein 1913, 1922; D. Bohm 1980, 1993; see also K. Wilber 1985), anthropologists (Gr. Bateson 1977, 1979), and musicologists (J.-E. Berendt 1985, R. Haase 1989). So far there does not seem to have been taken any appropriate steps to follow this fundamental principle in the practical application of virtual experience. The suggestopedic method of language assessment of G. Losanov (Ostrander/Schroeder 1979) made a radical step ahead in considering the balance of sensations and showed how "musical vibrations" can influence the sub-conscience to a relaxed, well concentrated recipient of language knowledge.

Summary and Conclusions

Based on the above critiques we can point out the following crucial issues, well known since Ancient Times,[67] but not consistently applied to the virtual reality implementations and their application, including (what is nowadays

[67] Heraclitus (Stobaios III 129 : Diels / Kranz 22B 112; Hippolytos Haer.IX 9,5 : Diels / Kranz 22B 51), Pythagoras, Plato (R.546A, Epin.991A, Ti.35A) and Aristotle (Cael.290b).

labelled as) the cyberspace:

- the fundamental ground of the natural feeling *inside the world* are *sound vibrations*;
- as long as the "*musical vibrations*; of the natural world are "integrated" just for advertising or serving to "*erotic*" purposes[68] we'll have only a constructed simulation of what an "*insider*" feels in a *virtual world*; and then comes a *second wrong step* trying to *reconstruct* some "hearing dimension" of ambient sounds, complementary to the visual framework.

We propose to approach this problem by introducing the *natural hearing* and *seeing dimension* to expand the *natural world* into the augmented reality design of virtual worlds. The guiding principles of such design are encoded in the experimental foundations of harmony, manifested in musical harmony, the harmony of the distances between planets and the order of the elements in Mendeleev's table (Shevelev et al. 1990). Our drive for imitation should lead us not to invent "what is natural," but to *hear it*. Because *hearing* is the starting point of *feeling* the "*whole" universe* (Bohm / Hiley 1993).

The clue of *opening* the "*pearly gates of cyberspace*" is not laying only in *approaching back* the historical experience of our civilization,[69] not in building up a symbolic lobby through a *discovery* of "*spiritual" dimension* in it, just to catch the attention of all spiritual religionists,[70] but in *trying* to *hear it* and to *feel it* among everyday life activities. Every beat of the heart, that could be the key of a musical success, every rhythm (and melody) that makes us relax, even the sound intervals of "silence" are part of this *wholeness* we are looking for. The variety of its appearance is the reason that it could be shown in so many different ways of evidence (orthodox choir opening the space at a click, baroque music etc.). For example, particular sound loops, derived from sections in orthodox choir chants, can be used as the mood setting *ambient sounds*. The cultural diversity in a global virtual world gives us the opportunity to personalise them referring to their religious and cultural background; so the original sounds can be filtered in a way that preserves the harmony, but changes the timbre.

The main problem in the experimental process to solve our dilemma is how to introduce *universal sound vibrations* (as developed by Kepler) and keep away

---

[68] Such as meant in Plato's dialogues (esp. Pl.Smp.201d-207a; see K. Dover 1980 : 1-2, 13-14) followed by others (M. Heim 1990, J. Lanier : Zaleski 1997 etc.).

[69] As done by M. Wertheim 1999.

[70] Here is a wide stream of healing therapies, religion sites (see Zaleski 1997) up to "confess online" etc. see esp. www.healthynewage.com/tsound.html.

from any mystical stream (Kabbalah, Tantra etc.) trying to "heal your soul."[71]

The practical answer is more than simple: the *musical intervals* have already found a "*virtual reality*" and we have only to *follow* "with the eyes" to the *binding field* of the *shared space*. The crucial point is to *bind* our emotions to the *natural concentration* of "*total listening*" of the natural feeling. If we do not concentrate emotionally (not only visually or rationally), we are going to fail in an *"emotional fallacy,"* seduced by the free flow of emotions. This step will cause us to "*paint*" an artificial recess for the happy few who would "*see*" and "*hear*" it; what is equally to an insider of a small group (called "universe," not a "pluriverse" in the sense of William James). Our aim here is to "create" *a shared room* for everyone willing and prepared to *hear it*, not a virtual reality for a "*social group*" (or a small elite).

References

Aristotelis. 1871-73. *Opera*, ed. I. Bekker. Berlin.

Bateson, Gregory. 1977. *Steps to an Ecology of Mind.* San Francisco: Chanler.

—. 1979. *Mind and Nature: A Necessary Unity*. New York: Dutton.

Berendt, Joachim-Ernst. 1985. *Das dritte Ohr*. Hamburg.

Bly, Sara. 1982. Sound and computer information presentation. Unpubl. doctoral diss., University of California, Davis.

Bohm, David, 1980. *Wholiness and the Implicate Order*. London 1980.

Boh, David and B. J. Hiley. 1993. *The Undivided Universe.* London / N.York 1993.

Chen, C. 1999. *Information Visualization and Virtual Environments.* London 1999.

Deleuze, Gilles and Félix Guattari, Félix. *Mille plateaux*. Paris 1980.

Dover, Kenneth, ed. 1980. *Plato Symposium.* Cambridge University Press.

Einstein, Albert and Marcel Grossmann. 1913. *Entwurf einer verallgemeinerten Relativitätstheorie und einer Theorie der Gravitation*. Leipzig.

Einstein, Albert. 1922. *Grundzüge der Relativitätstheorie.* Leipzig.

Haase, Rudolf. 1989. *Johannes Keplers Weltharmonik, Der Mensch im Geflecht von Musik, Mathematik und Astronomie*. Ahlerstedt 1989.

Heim, Michel. 1991. The Erotic ontology of cyberspace. In *Cyberspace: First Steps,* ed. Michael Benedict, 59-80.

Heraclitus 1960. *Fragmente der Vorsokratiker (Griechisch und Deutsch),* eds. H. Diels and Walther Kranz. 3 Bände. Berlin.

Kepleri, Joannis. 1940. Harmonice mundi, Libri V, MDCXIX. In *Gesammelte Werke,* Bd. VI, ed. Max Caspar. München.

[71] See his apology against the Oxford cabbalist R. Fludds 1622.

—. 1940. Mathematici pro suo opere harmonices mundi apologia, Adversus demonstrationem analyticam CL, V. D. Roberti Vluctibus Medici Oxoniensis, MDCXXII. In *Gesammelte Werke, Bd. VI,* ed. Max Caspar. München.

Krueger, Myron W. 1982. *Artificial Reality.*

Maher, M. L., S. J. Simoff, N. Gu, and K. H. Lau. 1999. Two Approaches to a Virtual Office Design, International On-line Conference on Design Computing DCNet'99, Sydney, http://www.arch.usyd.edu.au/kcdc/journal.

Ostrander, Sheila & Nancy, and Lynn Schroeder. 1979. *Leichter lernen ohne Stress (Superlearning), Die revolütionäre Losanow-Methode zur Steigerung von Wissen und Gedächtnis durch müheloses Lernen.* Bern.

Platonis. 1892. Dialogi, ed. J. Burnet. Oxford.

Qvortrup, Lars, ed. 2001. *Virtual Interaction: Interaction in Virtual Inhabited 3D Worlds.* London.

Rheingold, Howard. 1991. *Virtual Reality*. New York.

Sherman, William R. and Alan B. Craig. 2003. *Understanding Virtual Reality.* San Francisco 2003.

Shevelev, I. Sh., M. A. Marutaev, M. A. and I. A. Shmelev, I. A. 1990. *Section Divine, Three Views on the Nature of Harmony.* Moscow: Stroizdat. (in Russian).

Sutherland, Ivan E. 1963. Sketchpad: a Man-Machine Graphical Communication System. SJCC.

—. 1965. The ultimate display. *Proceedings of the IFIP Congress* 2: 506-508.

Wertheim, Margaret. 1999. The Pearly Gates of Cyberspace: A History of Space from Dante to the Internet. London.

Wilber, Ken, ed. 1985. *The Holographic Paradigm and Other Paradoxes.* Boston.

Zaleski, Jeff. 1997. *The Soul of Cyberspace.* San Francisco and New York.

CHAPTER FIFTEEN

THE DECONSTRUCTION OF PROGRAMMING, AND PROGRAMMING AS DECONSTRUCTION

SEUNGKWON JANG
DEPARTMENT OF LOGISTICS AND INFORMATION SYSTEM, SUNGKONGHOE UNIVERSITY

Program and Writing

In depicting the process of computer programming, the words "write" and "program" are invariably used together. "Program" as the object noun is always coupled with the verb "write." To "write" and to "program" are inseparable action words in programming. The definition of program is derived from "pro" (pre, before, forth) and "gram" (written mark, writing). Thus, "pro-gram" means "pre-writing."

In a similar vein, Derrida discusses the notion of program as a form of "writing." He addresses the etymological meaning of "program" in his book, *Of Grammatology* (1976, 9-10). And he asserts that biologists have been among the first to study "writing and program." Through "the processes of information in the living cell," writing becomes a program, which is embodied in DNA. In fact, DNA as a set of instructions in the living cell, tends to be called a program. In the wider terms of Derrida's conception of grammatology and writing, both the cybernetic program and DNA are seen as forms of writing.

Other applications of program reveal its function of structuring or arranging in sequence. For instance, the anatomist J. Z. Young (1978) uses the term "program" as a way of elucidating the functions and mechanisms of the brain. He argues that living organisms, especially animals, are governed by "sets of programs written in their genes and brains." Program in this sense is very similar to the technical terminology of the computer program (as the product of strong programming). Young (1978, 8), in fact, acknowledges the affinity between his analogy of program and the computer metaphor:

> I have in mind also the modern use of the word in "computer program," although

brain programs are only partly like the algorithms of computer software in which every step is logical.

How, then, do programs of the brain operate? In Young's argument, program is "written" and significantly by means of learning process that represents the phenomenon of living organisms. Young (1978: 10) argues that "each individual mammalian life follows a program written in two main languages (or four in man) embodied in distinct media": (1) The fundamental program of living organisms is written in the *DNA code*; (2) The second level of program is embodied in the *structure of the brain*; (3) The third level can be seen in *speech and culture*, in the organized sounds of spoken language; (4) The fourth level of program is found in *writing and other forms of recorded speech*. Young continues that the media of program consists of both biological and cultural media. The biological media of program are DNA and the brain, whereas the cultural media are speech and writing.

The etymological and empirical meanings of computer programming cannot be directly separated from writing. The etymological meaning of programming (i.e., "pre-writing") should be understood as "strong" programming in the sense that it means to prefix later happenings, and therefore to predict and control future events.

What, then, is writing? Is it simply the action of inscribing characters on the surface of paper? Obviously, writing began with the inscribing or carving of marks on some surface. *The Oxford English Dictionary* (2nd ed.) defines "to write" as: "to score; to wear by rubbing; to cut; to tear." Furthermore, these are historical and geographical variations of the word "write" in many European languages (see *The Encyclopaedia Britannica*). In English, to write and to incise; in German, *reissen* and *einritzen*, whose meanings are to wear and scratch; in Greek, *graphein*, which means to carve; in Gothic, *meljan*, to paint; in Slavic, *pisati*, also to paint. It is noteworthy that the origins of the word "write" in these various European languages come from the action of "carving," "painting," "inscribing," "scratching," etc. In short, the physical origin of writing is derived from the basic operation of "carving" (Coulmas 1989, 19).

The central theme of the paper will be the deconstruction of programming. Programming can be viewed as a representation of the "will to power" (and the "will to program"). In the present paper, the other nature, i.e., "weakness" of programming is explored in terms of writing and difference.

Viewing programming as writing creates an entirely new perspective on programming; it shows that "strong" programming has a "weak" shadow or underside, which is normally ignored. The weak-strong division reflects an earlier tradition in Eastern (Chinese) philosophy and thinking, namely, the Yin-Yang dynamism.

Yin and Yang are regarded as the two fundamentally opposing terms, which explain all natural and social phenomena. These two forces are generally contrasted (Needham 1956): Yin is associated with darkness, femininity, the maternal element, and Earth; Yang with light, masculinity, the paternal element, and Heaven. In terms of the Yang aspect, programming is viewed as a conventional tool of writing and inscription which facilitates order and control; that is, the strong aspect of programming. In contrast, the Yin side of programming is derived from writing as difference and separation, which leads to consider the weak aspect of writing. These opposing aspects of programming are described and determined by each other. In other words, the Yin side of programming, i.e., weak programming, undermines or deconstructs the strong aspect of writing while Yang seeks to assert its strong identity by resisting the weakening force of Yin.

Computer Programming as Writing

It is computer programming that represents the most technologically "advanced" form of writing. Computer programming is basically to do with manipulating logic, calculations and images, and communicating the others. This brings us to one of the main themes of this paper—computer programming as writing.

What are the specific advantages of viewing programming as a form of writing? First, it can serve to grasp the implications of programming in terms of a wider conceptual context. In this respect, computer programming is not only just an engineering practice, but it has an important meaning when seen in the context of the organizational and social worlds. Second, it is hoped that this paper can provide a good analogical perspective on writing and programming; also, it promises to offer a novel approach to the study of information technology in general. Thus, it may be possible to perceive more clearly recent technological developments in computer programming in terms of the broader implications of human writing systems. To do so, we have to answer certain questions about the notions of writing not only with regard to commonsense meanings (e.g., as a communication tool) but also from philosophical and sociological perspectives. Thus, these wider and deeper understandings of writing may illuminate the significance of computer programming as writing.

Writing is essentially a process of "scratching marks" on a physical space and medium (e.g., paper). With this observation, we can connect writing with the basic actions of the computer, as Palfreman and Swade (1991, 129) say:

> With a computer, numbers, letters and images take on electronic markings. Our ideas are represented as patterns of voltages in the circuits of a machine.

In the context of this perspective, we can grasp the conception of programming as writing that is as metaphor as well as physical activity. It is related to the meanings and roles of the computer and writing in the contemporary world. On the one hand, the computer is no longer simply a calculating machine; rather, it has become a medium for human communication. On the other hand, writing is not simply a means of communication in a literate society, but also is regarded as the general form of communication.

The role of the modern digital computer in society is often compared with the book. The modern computer can be better viewed as a "book" rather than as a "machine" (Palfreman and Swade 1991). We see more or less the same trend in the development of the modern computer, and specifically the personal computer (PC). Hence, both book and computer have become closer in terms of their functional roles in society. The point of this analogy between book and computer in the general context of our argument is that it serves to bring out the underlying similarities between the two as forms of writing and informational representation.

Studying the recent development of hypertext (textual networks for writing), Bolter (1991) insists on viewing the "computer as a new writing space," and argues that computing is writing. The writing space is a metaphor in the same sense that the writer can regard the mind as a writing space. Bolter's (1991, 9) perspective on the computer starts from the statement that "the computer is best understood as a new technology for writing." These arguments help us to understand the computer as writing space and programming as writing. Since computer programming can be seen as a form of "writing," it follows that computer programs can be viewed as "texts."

Bolter (1991, 10-11) further insists that "writing is the play of signs;" specifically, the computer offers a "new field for the play" of signs, and a "new surface for recording and presenting text." The surface that the computer provides is quite different from the conventional one. In this regard, the computer unfolds uncharted territory. In the traditional sense of writing, writing space is a kind of two-dimensional surface like paper. More significantly, the conceptual space of traditional writing (e.g., printed books) is "stable" and "monumental," which means that it cannot easily be changed or modified. Printed papers represent a fixed and rigid space; if someone tries to change printed material, it can be easily detected due to the apparent markings on the surface. This stable surface is related to strong programming whose function is to fix the fluctuating and fluid aspects of human activity. However, in electronic (computer) writing space, the most important feature is the "fluidity" of the space and the way in which this flowing space makes it possible to render an "interactive relationship between writer and reader." At this point, computer programming also has a weak side, which is the intention of this paper to explore in more detail.

The affinities between programming and writing can be found in the early history of the modern computer—*Turing test or game*. Alan Turing was a prominent figure in the development of Artificial Intelligence (AI). To introduce his ideas on AI, it is helpful here to describe the *Turing game,* or *Turing test*. Turing game is called the "imitation game" (Turing 1950). This is essentially a "writing game." The players of the game (that is to say, A (male), B (female) and C (interrogator)) ask questions and answer each other while they occupy separate rooms so that they cannot see each other. The final object of the game is for C as interrogator to correctly identify the two other (A and B) players' sex. Either A or B can be the computer. Turing intends that if human players (e.g., C) cannot tell the sexual identity of players (e.g., A and B), then the computer (either A or B) can be called a "thinking machine." This game can be said to be simply a game of "writing and reading" in the sense that it is totally concerned with the writing ability of the computer. As Bolter (1991) clearly points out, passing the Turing game means to be able to read and write (therefore to think), but has nothing to do with other faculties of being human such as touching, tasting, smelling, etc.

The word "writing" in its basic, commonsense meaning refers to the system of inscribing characters (such as the alphabet) on a surface (such as paper) in order to record or communicate information. But, the idea of writing in the context of this paper has a much wider meaning. For instance, Cooper (1989, 484) says: "writing is the process by which human agents inscribe organization and order on their environments." Thus, writing can be seen as a universal technology of ordering in human societies. Writing deals with the "structure and organization of representations," that is, "ordering, listing, display, hierarchy of arrangement, edge and margin, sectioning, spacing, contrasts" (McArthur 1986, 23). The term "writing" is much more than its limited, everyday meaning suggests—the functional writing of messages and their communication. Here, we intend the term "writing" as inscription in general.

The History of Writing

Tracing the historical development of writing shows us in what sense and to what extent programming itself may be considered as writing. In his book, *A Study of Writing*, Gelb (1963) discusses the evolution of writing. He begins with a tentative definition of writing: writing is "a system of human intercommunication by means of conventional visible marks." Later, Gelb (1980, 22) offers a more definitive characterization: "writing in its broadest sense is a recording system or device by means of conventional markings or shapes or colour of objects, achieved by motor action of the hand of an individual and received visually by another." In the context of these definitions, Gelb (1963) describes the evolutionary phases of writing. There have been, he says, roughly three stages in the history of writing; namely, primitive drawings, forerunners of writing, and full writing.

The stage of full writing is "phonographic" writing, which is concerned with "sound" and its inscription. Among current phonographic writing, alphabetic writing systems are the most widely practiced, covering almost the whole of the modern "literate" world. But the alphabetic writing system has a relatively short history. The first alphabet was probably invented by the Greeks (Gelb 1963). Most modern writing systems have followed the alphabetic system. The alphabetic system is the basic system for computer programming; indeed, most current programming languages are based on alphabetic systems, notably English.

Gelb (1963) insists that the "phonetization" (i.e., the transition from pictorial icon to phonetic symbol) is the most important single step in the history of writing. By virtue of phonetization, a primitive logographic writing was turned into a full writing system (Coulmas 1989; Gelb 1963). Thus, Gelb (1963) maintains that since phonetization has progressed, writing has lost its status as an "independent mode of expressing ideas" (through Chinese writing still has this property to some extent). Writing has become just a "tool of speech" and a "vehicle" in order to "scribe" exact speech and to be able to "record" information in a permanent and durable form.

Writing as Technology

Cooper (1989) views writing as the process by which organizations impose order both on themselves and on their environments. Writing is a technology that enables the "structure and organization of representation;" thus, writing is seen as a "way of fixing the flux and flow of the world." These statements represent "organization as a kind of writing" (Cooper and Fox 1989). In fact, it can be said that writing transforms "disorganization" into "organization;" these processes of organization and disorganization come together, and are different faces of the same phenomenon.

In a similar vein, Ong (1982) argues that "writing is a technology" in the sense that it depends on "the use of tools and other equipment" such as pens, brushes, papers, ink, etc. By using the term "technology," Ong wishes to underline a main feature of writing, "artificiality," by which he means that writing constructs human thought not by means of internal and natural (human or biological) ways but by means of external and artificial techniques. Similarly, Bolter (1991, 33) maintains that "writing is a technology for collective memory, for preserving and passing on human experience." In this way, writing facilitates and enhances "human capacity." Historically, writing has been the source of social order and organization, the very foundations of human civilization.

In contrast, as Ong (1986) indicates, writing has been criticized for its negative role in human thinking; for instance, Plato pointed out that the human memory capacity could be eroded by its dependence on writing. And more recently, a similar argument has been voiced to the effect that computers (or electronic calculators) make children lazy and incapable of calculating with their own unaided faculties. Because these technologies—writing and computers—are characterized as external and artificial, they are sometimes thought to degrade human thinking and creativity. But this understanding of technology derives from a thinking that views technology as a separate force from human beings and human action. If we view technology from a different aspect of thinking, technology and humanity are parts of an inseparable process. Bateson's example of the blind-person's walking stick is such an example of "integral" technology—it is an extension of the human body through which the blind person "feels" his/her way about (Bateson 1982). This is exactly the way that we approach writing technology in this paper -- it simply connects the human actor to its world.

Writing as Différance

Writing can be viewed in terms of strong sense of programming as well as weak sense of programming. However, as we have seen from the work of Coulmas (1989), writing inclines to strong facet of programming. One special feature of writing inclines it to "strong" facet—its tendency to separate itself from its source in the human agent. In terms of the separation process, writing can be seen to provide "strong" bases for understanding programming as writing in the sense of Ong's (1986, 36) characterization:

> One of the most generalizable effects of writing is separation. ... It divides and distances, and it divides and distances all sorts of things in all sorts of ways.

Ong (1986) suggests several facets in the separation process of writing: the known from the knower; interpretation from data; the word from sound; the source of the communication (i.e., the writer) from the recipient (i.e., the reader); the word from the existence; past from present; administration from other types of social activities; logic (thought structure of discourse) from rhetoric (socially effective discourse); academic learning from wisdom; being from time, etc. Among these, the foremost factor of writing in terms of separation is deemed to be the division between the known and the knower, the step by which writing promotes "objectivity" (Ong 1986). Likewise, naming something is a typical example of separation of the known and the knower;

naming the tree makes it possible to distance the known object (i.e., the tree) from the knower (i.e., the human perceiver).

Since Greek was invented, the "alphabet" system has been successful in the sense that it transforms sounds (spoken words) into their "spatial equivalents" (visual scripts). Translating phonetic marks and scripts from sounds means objectifying and formalizing the known which formerly was an inseparable and organic part of the knower's very being. These processes of separation highlight the features of writing as strong facet of programming, which create strong hierarchical ordering such as the known and the knower.

Ong (1986) also points out the relationship between formal logic and writing: writing separates "logic" from "rhetoric." Formal logic emerged from the Greek tradition because the invention of logic is intrinsically tied into the vocalic phonetic alphabetic writing system. The alphabet system is the most abstract form of writing in the sense that it may divide or distance more obviously and effectively and hence produce more seemingly objective and abstract forms, as we have already intimated. In this way, the alphabet system becomes an abstract writing system. One result of this separation process is that the conventional wisdom of the oral-based society has been superceded by "academic learning" or formal logic in which writing produces "highly organized abstract thought" that become detached (i.e., separated) from the more organic and "natural" activities of everyday life. In fact, so-called "scientific thinking" based on formal and abstract writing is unknown in oral-based society; writing makes it possible to think 'scientifically' because logical explanation is based on writing.

But separation can also be seen differently—not just as simple, static division but, more actively, as a process that also joins (Cooper,1986). If writing as simple separation helps us to grasp strong programming, the separation/joining structure helps us to see writing in a radically different way. "Writing" in weak sense is what Derrida's concept of *"différance"* deals with. Derrida has coined the term *"différance"* in order to reveal human ontology as "process." This can be otherwise expressed in Culler's (1979, 165) words:

> *Différance* ... designates both a passive difference already in existence as the condition of signification and an act of differing or deferring which produces differences.

This statement stresses the significance of the processual aspect of *"différance."* Derrida (1981, 38-39) himself maintains that *différance* should be both a "structure" and a "movement;" it is "the systematic play of differences, of traces of differences, of the spacing *[espacement]."* In this sense, Derrida's neologism *"différance"* is another name for writing but one which emphasizes the latter's endless motion and fluidity. Here, the motion and fluidity of writing as *différance* is equivalent to the denial of static separation and represents the affirmation of separation as temporary, active difference which seeks to join up with what it has been separated from.

Derrida's concept of writing starts from a criticism of "phonocentrism," by which he means that speech is given priority over writing itself (Cooper, 1989). Thus, Derrida tries to deconstruct this phonocentric hierarchy; moreover, he believes that writing has played a fundamental role in structuring human actions and social activities from ancient times. To express the significance of writing in human life, Cooper (1989, 486) uses an analogy to convey the Derridean notion of "writing:" "we are driven as we drive or, as Derrida would say, we are written as we write." In other words, human activities are essentially written or reflected by the act of "writing," an insight which contrasts with traditional thinking that writing is a tool for human communication or a mere transcription of phonetic signs. Thus, writing is not just a "supplement" of speech. Here, again, we are reminded of the weak (non-functional, complex and autonomous) character of writing, which leads nowhere except to its own realization.

Writing as the making of differences underlines the weak aspect of writing. Although writing is functionally dedicated to separation, and therefore to strong meaning of programming, programming as weak meaning of writing makes static division unstable and undecidable.

The Deconstruction of Computer Programming

Writing can be understood as *différance*; in other words, "writing is always an act of deferral" (Bolter 1991, 188). One simple example is found in the relations between writers and their words. A writer writes words (script) down on a writing surface (paper) so as to use and have those words "at a later time," which is a deferral effect. Following Derridean logic, several deconstruction analyses have been applied to computer programming. Programming as "electronic writing" institutes its own kind of *différance* in the way in which the computer is able to "intervene between the author and his or her words or between the reader and the author's words" (Bolter 1991, 189). In the same way, the computer programming is certainly able to make differences in the relationship between the author (i.e., programmers) and the text (programs).

Furthermore, the computer programming as electronic writing promotes the possibility of deconstruction practice more readily than does the traditional writing of graphic marks (Poster 1990). Electronic writing on the computer produces a new configuration of author and text. This change is perhaps most notably seen in the "subjectivity" that is constituted not by a "logocentric presence" (e.g., "speech") but by a form of "dispersion of the subject" (i.e., the "text" of electronic writing). As Poster (1990, 123) says:

> The electronic mark, as opposed to the graphic mark, permits a deep reforming of the space/time coordinates of writing and reading through everyday life. The electronic mark radicalized the anti-logocentric tendencies that deconstruction argues are inherent in all writing. The change from graphic to electronic mark permits the wide diffusion of electronic writing throughout the social space and undermines the temporal limits of pre-electronic writing.

In his work, *The Post Card*, Derrida (1987) chooses a specific picture to illuminate this new configuration of subjectivity and the deconstruction of the logocentric perspective on writing. The picture depicts Socrates writing at Plato dictation. This is exactly the opposite image to the conventional interpretation of the relationship between the two philosophers. In fact, historically it is regarded as "fact" that Plato wrote what Socrates said. Nonetheless, according to Derrida's logic of undecidability and deconstruction, Plato precedes Socrates as writing precedes speech, and it is this that the picture (actually a postcard seen in the Bodleian library at Oxford) of Plato and Socrates addresses.

Derrida uses the postcard as a novel way to present his argument. As Poster (1990, 126) describes it, the postcard can represent "a fragmented and disoriented subject" due to its openness and condensed expression. For these reasons, it is said that the postcard is an unstable medium. Moreover, since the postcard has the similar effects as writing (i.e., "unstable" *différance*), that is to say multiplication and decentering subject, the feature of the postcard shows how writing becomes the process of *différance*. This destabilization effect of the postcard is the same as writing, especially electronic writing and communication. The logic of fragmentation and disorientation is seen not only in postcard, but in electronic writing (Poster 1990).

With regard to its multiplication and instability, computer programming can be paralleled with writing in general. But we may see the similarities and differences between traditional writing and electronic writing. On the one hand, new technology of writing such as computer programming "tends to be absorbed within the category of writing" (Poster 1990, 127); it means computer writing (programming or tele-writing) becomes writing in general. On the other hand, computer writing is different from traditional writing. While conventional writing locates itself in linear space and temporality, computer writing "inserts itself in a nonlinear temporality that unsettles the relation to the writing subject" (Poster 1990, 128).

The introduction of new writing, i.e., electronic writing, may have a similar impact on the configurations of the self, the human actor and time/space, etc. It has been said that the Industrial Revolution created widespread mobility through the technology of transportation; this made the "dispersal of the self" a possibility. In other words, modern transportation challenged the old barriers in time and space. Likewise, it can be argued that computer writing "intensifies" and "radicalizes" this tendency to "scatter" the human subject in "nonlinear spatio-temporality." The computer, or electronic writing, promotes "immateriality;" and it brings with it the "disruption of stable identity."

Using Derridean notions, Poster (1990) has discussed the nature of electronic writing, including computer programming. He argues that Derrida's idea of writing can be viewed from two positions: Derrida's early position on the "relation of the subject to writing" (for instance, in *Of Grammatology*) and the later position on writing "as a theoretical strategy for the analysis of computer writing" (notably, in *The Post Card*). However, both theoretical stances are by and large equally significant in the sense that Derrida's term "writing" has to be considered as "always already anterior to speech." Furthermore, "writing and speech perpetually "oscillate" so that each cannot be fixed in relation to the other" (Poster 1990, 103). This oscillating character of writing can efface the boundary between subject (e.g., author or programmer) and object (e.g., text or program). This undecidable feature of writing places electronic writing closer to the "border of subject and object." Unlike traditional writing which is located on a stable surface, computer writing always and already "dematerializes the written trace," that is, it creates immateriality. Thus major characteristics of electronic (computer) writing are said to be its movability and alterability, which make it instantly transformable and decomposable, i.e., immaterial. For such reasons, computer writing can be called a 'border line event' between "stability" and "instability," between "subjectivity" and "objectivity."

In addition, computer writing makes ambiguous conventional "Cartesian dualism" which represents the clearly and distinctively demarcated relationship between subject and object (Poster 1990, 112). For instance, conventionally, a manuscript in handwriting can be authorized as the "original" that makes us trace the "true" and "original" texts and the evolution of the text. However, computer writing can hinder the search for the "originality" of texts and the "authenticity" of authors due its tendency to "dematerialize" the text. For this reason, one can argue that the computer creates a collective authorship and so decenters the traditional subject of writing: "computer writing subverts the author as centered subject in yet another way, by introducing new possibilities of collective authorship" (Poster 1990, 108).

These issues of computer writing and collective authorship are reflected in many ways in recent developments in the computer field. For instance, "groupware" promotes the development and utilization of collabourative multi-user word-processors in place of the more traditional single-user word-processors. Poster argues that individualistic authorship gives way to the "indeterminate author." Computer technology erases the author's voice and trace the ways in which author's own manuscript cannot be kept any longer in conventional places (e.g., museum) and conventional forms (e.g., manuscript). Furthermore, computer programming undermines the "stability of the position of the author." The computer makes it easy to search for "supplementary authority." As Poster (1990, 115) puts it:

> In sum, while computer writing dematerializes the mark, merging in a new simulation of unity the written object with the writing subject, it also subverts the individuality of the writer in the marks it does leave, the files on the floppy disks. Finally, it creates new possibilities of collective authorship.

We can instance various examples of this tendency; e.g., computer-mediated communication and computer conferencing. A new form of communication, e-mail becomes a "new configuration" in the constituting process of self. This is more dramatically represented by computer conferencing, which can undermine traditional ways of thinking such as the "Cartesian subject and social hierarchies" because in computer conferencing communicators can hide their social origins and hierarchical position in organizations (e.g., accent and behavioural pattern) and therefore create a kind of collective anonymity. Thus, the hierarchies between communicators in conventional communication systems become obsolete. Similarly, writing on the computer, especially in computer conferencing, can introduce a new relationship between author and text. In computer conferencing, we observe a form of dispersion of the subject in terms of space as well as time. The subject is no longer stable and identifiable in this system. In terms of space, the subject is not present in the same place over time: authors become dispersed throughout networks of communication. Hence, the tendency for computerization decenters the conventional subject-agent in a collective anonymity.

Time and Space in Electronic Writing

Many of the issues raised above in connection with computer writing are bound with the changes that the new communication technologies create in our experience of time and space. In fact, computerization of time and space has become one of the more commented-on features of advanced communication technology. Notably, Lyotard (1991) has argued that information and communication technology should be understood in terms of difference and memory in the constitution of time and space. Specifically, Lyotard introduces a number of neologisms to illustrate the novel significance of the new technologies i.e., the computer and related technologies: tele-graphy, tele-writing, inscription from afar, writing at a distance, inscription-at-a-distance, etc.

Lyotard's point is to underline the implications of the new communication technologies on the way we experience space and time. His argument starts with the writing (or inscription) model of technology, and this produces a "spatialization of meaning." Partly due to its legibility or decodability, and partly due to its marking on a spatial surface, inscription (writing) can open a "public space of meaning" (e.g., books and newspapers) and produce a "community of users-producers" (e.g., writers and readers).

In the context of this spatialization, inscription produces and maintains the communal memory. Likewise, tele-writing keeps the "sign of the past event" so that it generates "available, presentable and reactualizable memory." Lyotard also argues that graphics are a form of inscription. Representing a form of spatial memory, inscription means that the "thing can pass" but the "sign remains there." As a kind of mnemotechnic, graphic inscriptions conserve engraved traces just like the museum conserves an ancient artifact. In other words, inscription can be said to be "memorization or conservation." In this sense, inscription "conserves" time by keeping the past in the present.

As Ong (1986) already noted, taking "being" out of "time" might be the "most momentous" of all "effects" of writing (inscription) by which the time dimension should be eliminated from being by writing. With the writing effect (i.e., separation being from time), "becoming becomes being" (Ong 1986, 44). The separation "past" from "present" and the creation of "distanced past" might be the expressions for illustrating that segregation effects of writing are abolishing the time-dimension in many of social and historical discourses. Changing and "mobile" oral world (i.e., becoming) have been replaced by stable and constant "text" (i.e., being) which derives from writing.

As inscription devices, the computer and related new technologies promote "mental operations" such as memorization, regulation, and calculation. Relating to these roles, the computer has impacted another way in social life, which is transformation and communication through time and space. In the computer, the digitalized data can be synthesized in any time and any places. These are "rendered independent of the place and time of their "initial" reception, realizable at a spatial and temporal distance" (Lyotard 1991, 50). By Lyotard, it is called "telegraphable" which points out "noteworthy facts" concerning memory breaching features of tele-writing. For instance, by means of specific mode of tele-graphy (e.g., networking computer programs), writing at a distance (i.e., information communication technologies) removes the social contexts and social formation of hierarchies in communication (see the previous section of this paper). These social memories (i.e., social context and class) are diminished by tele-writing.

The computer and inscription-at-a-distance can break the traditional relations of time and space in terms of memory and remembrance. Since conventional thinking views time and space as stable and unitary dimensions as if these are based on material foundations, strong facet of programming may support stable and unilateral features of time/space. But the weak facet of programming of tele-writing can break the rule of unchangeability and stability in time/space dimension. Thus weak meaning of programming is sensitive to the precarious time/space dimensions.

In conclusion, we have tried to show that programming can be deconstructed through poststructuralist thinking. Writing as viewed through Derrida's concept of différance provides the key to the deconstruction of programming.

References

Bateson, G. 1972. *Steps to an Ecology of Mind.* London: Intertext Books.

Bolter, J. D. 1991. *Writing Space: The Computer, Hypertext, and the History of Writing*. Hillsdale, NJ: Lawrence Erlbaum.

Cooper, R. 1986. Organization/Disorganization. *Social Science Information* 25.2: 299-335.

—. 1989. Modernism, postmodernism and organizational analysis 3: the contribution of Jacques Derrida. *Organization Studies* 10.4: 479-502.

Cooper, R. and S. Fox. 1989. Two modes of organization. In *Frontiers of Management,* ed. R. Mansfield, 247-261. London: Routledge.

Coulmas, F. 1989. *The Writing Systems of the World.* Oxford: Basil Blackwell.

Culler, J. 1979. Jacques Derrida. In *Structuralism and Since: From Levi Strauss to Derrida,* ed. J. Sturrock, 154-180. Oxford: Oxford University Press.

Derrida, J. 1976. *Of Grammatology*. Trans. G. C. Spivak. Baltimore: The Johns Hopkins University Press.

—. 1981. *Positions*, trans. A. Bass, A. Chicago: The University of Chicago Press.

—. 1987. *The Post Card: From Socrates to Freud and Beyond.* Trans. A. Bass. Chicago: The University of Chicago Press.

McArthur, T. 1986. *Worlds of Reference.* Cambridge: Cambridge University Press.

Gelb, I. J. 1963. *A Study of Writing*. Revised ed. Chicago: The University of Chicago Press.

—. 1980. Principles of Writing Systems within the Frame of Visual Communication. In *Processing of Visual Language 2,* eds. P. A. Kolers, M. E. Wrolstad, and H. Bouma, H., 7-24. New York: Plenum Press.

Lyotard, J.-F. 1991. *The Inhuman: Reflection on Time*, Trans. G. Bennington and R. Bowlby, R. Cambridge: Polity.

Needham, J. 1956. *Science and Civilisation in China: Vol.2 History of Scientific Thought*, Cambridge: Cambridge University Press.

Ong, W. J. 1982. *Orality and Literacy: The Technologizing of the Word.*

London: Methuen.
—. 1986. Writing is a technology that restructures thought. In *The Written Word: Literacy in Transition*, ed. G. Baumann, 27-30. Oxford: Clarendon Press.
Palfreman, J. and D. Swade, D. 1991. *The Dream Machine: Exploring the Computer Age.* London: BBC.
Poster, M. 1990. *The Mode of Information: Poststructuralism and Social Context.* Cambridge: Polity.
Turing, A. 1950. Computing Machinery and Intelligence. *Mind* 59.236: 433-460.
Young, J. Z. 1978. *Programs of the Brain: Based on the Gifford Lectures, 1975-7.* Oxford: Oxford University Press.

CHAPTER SIXTEEN

TEACHING WITH TIKI*

MICHAEL BYRON
KENT STATE UNIVERSITY

Introduction

Many philosophy instructors came up in departments that taught in a traditional style. Our professors lectured to us while we took notes, they encouraged us to ask questions when we sought clarity on a particular point or passage, and required us to write a term paper. The model worked for graduate school, too. Some instructors encouraged more discussion in class than others. Work outside class meant reading the assigned text, talking about it with a friend, and writing the term paper on your own in the library.

Philosophy instructors today are looking for ways to expand this pedagogy, with its top-down emphasis and its limited modes of interaction. Many of us want our students to do more than just read the text and take notes in class: we would like them to discuss the material, to work on collabourative projects, to think and talk about the material outside of class, and to make our course material part of themselves. We also wish to take advantage of technology, though in ways that support, not dominate, our teaching.

At Kent State University, I have worked on implementing technological enhancements of the traditional pedagogical model. The particular tool I would like to share with you is a computer application called TikiWiki, or Tiki for short. I have used this software to create an entire website environment for my philosophy classes, and this environment has allowed my students to learn the course content in ways quite different from the traditional model: the learning is different because (as Dewey emphasized) we learn what we *do*. This article describes the Tiki, presents some considerations in favor of deploying it in philosophy courses of very different kinds, and explains how I have used the

* This article has already been published in *Teaching Philosophy.* I would like to thank Prof. Michael Goldman, the editor, for giving permission to publish it in this volume.

software in two of my courses.

What is Tiki?

Tiki is a web-based content management system (CMS), and it can be downloaded over the web at http://tikiwiki.org. When installed on your server, it provides a variety of tools, including discussion forums, messaging, webmail (but not a mail server), chat, file galleries, image galleries, blogs, and literally dozens more.[72] All of these features are deactivated by default: when you set up Tiki you activate whichever features you wish to use via a straightforward administration interface.

Many of these tools will be familiar to readers: by now most of us have heard of if not used email, internet messaging, chat rooms, discussion forums, and file galleries (or FTP spaces). Perhaps the most distinctive and least familiar tool provided by Tiki is the wiki. A wiki is a collabourative website, based on the original WikiWikiWeb at http://c2.com/cgi/wiki and deployed brilliantly in the WikiPedia (http://en.wikipedia.org). It provides a simple, form-based way to create and edit web pages right over the web, requiring no knowledge of HTML or any significant markup language.[73] Users visit a page, click "edit," type the text they want into the web form, and click "save." Their content is saved as a new or modified webpage. No muss, no fuss, no HTML. The wiki is great for collabourative class projects, as it allows students to meet virtually at their convenience and work on documents together.[74]

---

[72] For those not familiar with these web services: discussion forums are like electronic bulletin boards, where users can post and reply to messages related to a particular topic or range of topics. Internet messaging is a means of sending text messages over the internet instantly and conversing with text. Webmail is a web service connecting a website to an email server, so that users can read and send email without a separate email program or application. Chat is similar to internet messaging but allows for multiple users in addition to one-on-one communication. File galleries are internet sites containing files for download, and to which users can upload files of various kinds. Image galleries are a special kind of file gallery dedicated to uploading and downloading digital images. Blogs, an elision of "web logs," are electronic journals or diaries that are open to some community or the public.

[73] I say "significant" markup language because Tiki does allow some formatting in the wiki. For example, to create bold text, one uses underscores on either side of the bold text, _like this_. Similarly simple markup allows one to create headlines, horizontal rules, tables, and other standard web page formatting. The point I wish to emphasize, however, is that one need not know any of this to create a web page: one could use plain text. This simplicity is in marked contrast with HTML and other markup languages.

[74] I have used the tool to collabourate with colleagues at other institutions as well. For example, I set up a secure wiki in order to facilitate a State of Ohio Board of Regents

Many instructors will be familiar with CMS in the form of WebCT, Vista, or other applications deployed on school servers. At this point, those systems offer a fine interface and good support. As far as I know, however, none of the commercial products incorporates all of the features of Tiki. Of especial note is the absence of a wiki from most types of CMS. No other tool allows the collabourative development of webpages, and Tiki's inclusion of wiki functionality makes it a terrific tool.

The disadvantages of using Tiki depend on who installs it and how it is installed. It is open-source software, so it is costless in that regard. If a departmental server administrator who is familiar with software installations sets up and maintains the software, the process is fairly straightforward. Although Tiki can be set up to run with LDAP authentication (which allows students to use their university user ID and password), I have not done that; students thus need separate accounts and passwords for my Tiki. I discuss specific technology issues pertaining to Tiki at the end.

Why Teach with Tiki?

The dimension of teaching with Tiki from which my classes have benefited most is that it facilitates learning beyond the classroom. As an all-inclusive site, it has all the course content students need: information about the course—including syllabus, reading assignments, homework, and deadlines—as well as handouts, lecture notes, reading notes, discussion forums, and chat capability. All of these features are available via the internet all the time, 24/7.[75]

The communication fostered by using Tiki can be especially important in certain contexts. Three years ago I began teaching our symbolic logic course via distance learning, delivering the course using a synchronous (that is, simultaneous) computer-based format to as many as six campuses at once. Included in that course were three sections at my campus, in three different classrooms. Students at the regional campuses, though glad to finally have access to a non-math course that fulfilled their math requirement, expressed concern that they did not have good communication with me or the class. With as many as 100 students enrolled, but sometimes only five in the lab at a

project concerning transfer students at Ohio state colleges and universities. My colleagues across the state could log into the site and read, update, and edit the documents that the panel was working on. The site was a great convenience and saved us the trouble of emailing multiple copies of the documents around each time one of us edited them.

[75] To view the Kent State Philosophy Department's Tiki, please visit: http://philosophy.kent.edu/tiki.

regional campus, they felt cut off and isolated.[76]

I took two steps to address this concern. First, I initiated "online office hours" using chat functionality similar to that provided by Tiki. Since it was not feasible for most regional campus students to travel to my office, and since many of them are reluctant to telephone (and incur long distance charges), using technology this way seemed a fine solution. Many students took advantage of this form of access.

Second, I provided an online discussion board and made contribution to the board a requirement for the class (just one post per week, on average). This feature of the course took off immediately and was a terrific success. Students found that they could use the board to ask questions, to discuss issues in the reading, to seek help with specific logic exercises, to vent about the difficulty of the material, and to explain their own approaches to others in the class. Some even brought up current events to discuss, though I made clear that such posts did not count as a contribution to the class.

When I introduced these elements into my distance learning sections of logic, I was not yet using Tiki, which had yet to be developed. I found separate pieces of software that were not an integrated solution, the way Tiki is. I began using Tiki during the 2003-04 academic year, thus obviating the need to run all of the other services on my server.

One additional advantage of using Tiki bears mention. Many schools are pushing instructors to use more instructional technology, since colleges and universities are spending considerable sums on the equipment and support infrastructure for the classroom. Philosophy instructors sometimes resist these changes, insisting that our classes still work best with good, old-fashioned classroom discussion. I agree; but we can nonetheless use technology to extend that discussion beyond the classroom. I use Tiki to supplement, not to replace traditional classroom interaction, though in my distance learning logic class that interaction has already been displaced. Using one, or a few, or a dozen of Tiki's features can be an appropriate way for philosophy instructors to address the pressure to use more instructional technology.

Logic with Tiki

For the past 3 years, I have taught a section of our department's symbolic logic course via distance learning. Our specific setup has students at three to six campuses, stationed in front of computers. Each workstation receives an audio-video feed from me, and I can activate any student's camera at any time. One

---

[76] I have discussed the preparations for this project in *Teaching Philosophy* 24 (2001), 255-277.

person at a time is in the video window, though when students are there my microphone is also active, so that the class can see the student and hear our conversations. Their computers are also configured to share my logic software screen, so that everyone can see the text, examples, and exercises that we are discussing. The situation is quite different from the traditional classroom setting, and it took me several semesters to become accustomed to working with the system. But the distance learning classes are popular at the regional campuses, and the logic class is the only one I would consider teaching via distance learning.[77]

During the past academic year, I have begun using Tiki with my distance learning logic class. Tiki has the advantage of combining a range of web services in one convenient and easily installed application. The first item I post on Tiki for logic is the syllabus. I do not print a paper syllabus for the course: although the distance learning support staff have arranged a system for the distribution of paper (syllabi, exams, and so on), I prefer where possible to use the website, which is easier to update. My syllabus is nothing special: it includes the standard logistics information, a course description, and a detailed account of how the students' work will be graded. On a separate page I have the course schedule, including reading assignments, homework assignments, and test dates.

One intriguing feature of these pages is that they are posted using the wiki. That means that, with the permissions I have configured, any registered user can edit the content of these pages. You might suspect that some student would alter or deface the pages, but during the time I have used Tiki, this has not happened. Tiki keeps a record, of course, of who edits pages and when, and it keeps many (more than two dozen) previous versions of each page. So it is a simple matter to correct any vandalism and to discover the offending party. Moreover, checking for altered pages is simple: I click one link and see all changed pages for the past 24 hours, week, or month. So although the potential for abuse exists, I have yet to encounter it and will be able to correct it immediately.

A third section of the wiki contains course handouts. I have eight or ten handouts that I use during the semester, and I have pasted these into the wiki. One of the great features of using the wiki as opposed to other systems is that no conversion into HTML is required. The wiki editing page is a form window for plain text, and it is easy to cut and paste a handout from a word processor document into the form. Although state of the art word processors have an HTML conversion utility, the resulting pages have no return links and are web

[77] A complete discussion of the advantages and disadvantages of synchronous delivery of philosophy courses via distance learning is well beyond the scope of this article. Suffice to say that the philosophy courses other than logic that I teach are strongly grounded in class discussion and group collabouration. These modes of instruction are quite difficult if not entirely unfeasible given the technology in use in our distance learning labs.

dead ends. The handouts in Tiki are linked back to the rest of the logic site and to other areas of Tiki as well.

Fourth, I use Tiki's discussion forum for the logic class, asking that they contribute regularly, on average once each week. Here is an excerpt from my welcome message on the forum:

> Use the forum to ask or answer questions about the software, about logic, about philosophy, or anything else. Virtually any post relevant to the course will count, whether you're asking a question, answering a question, or making a comment.

As I have discussed, the forum provides a place for students to raise all sorts of concerns, mostly though not exclusively about the course or its instruction. (I have had to delete only one fraternity party announcement, sending a private message to the poster that such matters were not sufficiently related to the course.) The forum implementation in Tiki is not the best available; for example, I like phpBB better, and even the forum in WebCT is not bad.[78] Still, there is something to be said for an integrated suite of tools, such that only one login is required of the students.

Seminar with Tiki

The fit between the distance learning logic course and the use of a technology like Tiki was more obvious to me than fitting Tiki into a more traditional philosophy course. I had used course websites for my introductory classes, but they had been strictly informational. Those sites included syllabi, schedules, and lecture notes. The communication facilitated by those sites was strictly one-way, from instructor to students. The students would use the site by printing it out and stuffing it into their notebooks. This usage saved the department some money on copier expenses, but using technology in these limited ways was not enhancing my teaching. More was possible.

During 2003–04, I decided a more radical approach was in order, and so I sought to take advantage of more of Tiki's features. I taught an upper-division seminar in the philosophy of religion, and I structured many of the assignments around using the website. Still included in the site were the syllabus and course schedule. The students were still required to contribute to the forum once each week. But I expanded use of Tiki in two directions.

First was an assignment I called WikiNotes.[79] Twice during the semester,

---

[78] The features of phpBB that appeal to me include its powerful administration interface and its tools for sorting and organizing posts. Developing the comparison would be tangential.

[79] The page naming convention in a wiki is to capitalize the first letters of at least two

each student would take a turn writing a summary and posting it on the wiki. These pages provided a useful guide to the reading, as well as another measure for me of how well the students grasped the readings. Moreover, since the notes were posted on the wiki, the content could be edited by other students. Unfortunately, I did not observe other students correcting errors in the summaries. I know from class discussion that students read each other's notes, but they seemed unwilling to "poach" on other students' work.

When I use Tiki again, I will look for ways to make these pages more collabourative. Assigning a pair of students to a particular reading did not achieve this goal, as the students simply divided the pages of the reading and summarized it as two distinct assignments. One idea I have in this regard is to change the nature of the assignment. In other courses, I have assigned what I call "reply papers," where students select one specific point from the reading. They summarize this point in a paragraph or two, and then they reply to the point from their own perspective in a paragraph or two. They may agree or disagree, or they may simply comment or reflect on the passage. It is possible that this kind of assignment, when given to a pair of students, might generate an interesting page of notes for the readings.

The second direction in which I expanded the use of Tiki was for paper assignments. I decided to experiment with collabourative paper assignments. The collabouration took place at the editing, not the writing, stage of the assignment. Students would write their five-to-seven-page papers as usual, but instead of handing them in, they uploaded them to Tiki's file gallery. The gallery is similar to an FTP space, but easier to use and protected (only members of the class could access it). Next, they had to download another student's paper, read it, and make comments using Microsoft Word's "Track Changes" feature (the use of which we went over in class). For the purposes of guiding this exercise, I handed out a writing rubric that included a page full of tips for improving writing. We talked about how to use the rubric and how to make constructive comments: for example, asking questions ("What does this mean?") is more helpful and positive than making comments ("Vague"). Students would then upload the paper on which they had commented to the file gallery.

The last part of the assignment was for the students to download their own papers, on which someone else had commented, and to revise those papers in light of the comments. They then uploaded the final version to the file gallery for me to download and grade. I do not mind reading papers on the computer, so

---

words and then remove spaces to create a WikiWord (like that one); the software recognizes such formatting as designating page names and makes them into web links to the named pages: this convention is more cumbersome to describe than to use. Hence the term "WikiNotes" for the pages in my seminar's site.

I would grade them and email them back to the students (to protect the privacy of the grade). Other instructors might at this point ask students to print out their papers and turn them in at class.

The students had somewhat mixed reactions to these assignments. Naturally, many disliked the extra work. But most recognized how much better their final work product was than the first draft. If nothing else, the exercise might have taught them the value of editing their own work. I suspect, however, that they also gained some insight into how to edit and how to read student work critically. I was pleased overall with this kind of collabourative assignment and will use it again in the future.

I was less satisfied with the seminar's use of the discussion forum. Many of the students struggled to contribute once each week; some did not contribute at all. The standard complaint is that "I had nothing to say;" we have all heard this before. It might be that I simply asked too much of the students, that the reading assignments, WikiNotes, exams, and papers were sufficient, and that the forum was otiose. Still, I am reluctant to give up on it, providing as it does such a great way to extend class discussion. How many times have you had a student raise a startling and intelligent question at the end of class and wish you had an extra 15 minutes to explore it? The forum is an ideal solution, and I plan to use this idea in the future. In addition to asking students to post their end-of-class queries, I will "plant" some provocative questions and ask the students to respond. They will be free, of course, to introduce their own questions, either as tangents to mine or as separate discussion threads. But some focus for the discussion should facilitate more widespread contribution.

Technical, Legal, Pedagogical Issues

Let me conclude with some remarks about technical and other issues that arise through the use of Tiki. Perhaps the most important is security: with the number and severity of virus and other security problems on the internet today, many will be leery about installing a new service on a server, especially one that is relatively new and unknown. Without getting too technical, Tiki uses a scripting language that is extremely secure (PHP), and it is compatible with a range of different operating systems (Windows and UNIX) and databases (flat file or most types of SQL). Tiki is in active development by the open source community, however, and releases frequent updates, bug fixes, and security fixes. In other words, it keeps up as well as anything with the security issues of the internet.

A distinct issue concerns FERPA, the Family Education Rights and Privacy

Act.[80] This U.S. legislation protects the privacy of student education records, including work products and grades, and it presents a potential obstacle to any kind of written collabourative work in class. The issue is that under FERPA student work should not be distributed or exposed to anyone other than the instructor without the student's explicit consent. This would leave participation in collabourative work up to the individual student, and instructors could hardly assign it under those conditions.

Consequently, in order to use Tiki, I must require students in my classes to waive some of their rights under FERPA. I do this by including a disclaimer in the syllabus, which explains the situation and states that by remaining in the class, students agree to waive their privacy rights with respect to collabourative work for the course. The waiver is explicit and limited, and it does not extend, for example, to the students' grades for the course or any other evaluation by me.[81] I have discussed this issue and vetted my specific disclaimer with our school's director of network security, and this seems to be the best solution, given existing legislation.

Finally, as with any new technology, Tiki has a learning curve, both for instructor and student. The instructor must learn at least how to administer it and use it, if not also to download and install it. Students must learn how to use it as well, though my experience is that the vast majority of students are sufficiently web-savvy that their learning curve has a slope approaching zero. Tiki has documentation, both built-in and online, and the community of developers behind the application are helpful and respond quickly to inquiries. The benefits of teaching with Tiki easily repay the investment of time and other resources to implement it in the philosophy classroom.

---

[80] For more information about FERPA, please visit: http://www.ed.gov/policy/gen/guid/fpco/ferpa/index.html.

[81] Here is the specific notice I have employed on my syllabi:

FERPA Statement

This course may require contribution to the discussion forum, wiki, and other collabourative projects. Collabourative projects by their very nature expose student work products to other students and possibly members of the public, although graded work and student grades will never be disclosed. Nonetheless, the nature of collabourative work makes it impossible to protect all of students' privacy rights under applicable provisions of FERPA (Family Educational Rights and Privacy Act). Students remaining in the class thus agree to waive their FERPA rights for any collabourative work for this course.

CHAPTER SEVENTEEN

STUDENTS' EXPERIENCE ONLINE LEARNING IN THAILAND

PANNEE SUANPANG

RAJABHAT SUAN DUSIT UNIVERSITY

Introduction

In the past decade, Information and Communications Technology (ICT) has developed its potential to enable national economic, social and educational development. Online learning has been promoted as being more cost effective, more convenient, and providing increased opportunities to the life-long learner (Olson and Wisher 2002; Richardson and Swan 2003). It has demonstrated several advantages over a traditional learning environment, particularly in the concept of "learning anytime and anywhere". This means that student access to online course material is time and place independent (Berge 1997; Matthews 1999).

In Thailand, the government embarked upon substantial educational reform with the 1999 National Education Act (NEA). The key aspects of this reform focus on improving efficiency and effectiveness of learning. Students were encouraged to become critical and creative thinkers, to acquire facility with information technologies, and to develop their learning and individual potential based on the philosophy of "student-centred" learning. In 2002, the Thai government announced plans to install computers connected to the Internet in all high schools, and to make the Internet and "E-learning" or "Online learning" the technology of choice for the Thai higher education system.

The Rajabhat Institutes form a nation-wide network in Thailand with a uniform approach to teaching and learning. Instructional materials for 11 general educational courses were produced using an E-learning model. Recently, the Suan Dusit Rajabhat University (SDU) in Thailand has encouraged the development of effective and efficient course materials for delivery through Internet technologies, in order to increase and enhance opportunities for both campus-based and distance learning. The Institute's goal is to become a "Cyber University" in the next five years. A preliminary investigation of current problems with students' learning in Business Statistics found that there were high failure rates, with most students feeling that it was difficult to understand statistics concepts and to apply them to the real world. Missed lectures were a particular problem, as their understanding was impaired when they returned to their studies and catching up became very difficult. This unsatisfactory experience with Business Statistics highlighted the need for a new teaching strategy to improve its outcomes, so it was decided to trial a pioneering study of the effectiveness of an online course in Business Statistics in Thailand.

The purpose of this paper is to explore students' experiences in learning in an online Business Statistics course and to examine the qualitative reflections that they provided throughout the course and as part of the course evaluation. This exploration of student experience is focused on these questions:

- How do students experience learning Business Statistics in both traditional and online methods?
- What are they learning and gaining from the course?
- What features of the course do they like or dislike, and what suggestions do they have for change?

Literature Review

Higher education around the world is facing challenges to meet new demands for utilizing information technology to improve quality of learning. Various criticisms of traditional classrooms appear frequently, such as lack of teacher attention, boredom, outdated knowledge, and inappropriateness for a diverse population (Gardiner 1997; Hara & Kling 1999). Many researchers advocate new concepts such as active learning, student-centered principles, effective use of technology, and collabourative learning (American Psychological Association 1997; Bonk and Kim 1998). There are high expectations that technology will transform higher education. This demand has also impacted on developing countries in South East Asia, such as Thailand. Therefore, it is important and timely for educators to explore students' attitudes towards and experience of new technology such as online learning.

the key features of online learning are time-independence, text format, computer mediation, multiple threads of conversation and fluid participation patterns. Online learning design tends to be based on constructivist and learner-centred principles that increase learners' control, facilitate the sharing of multiple perspectives, and encourage individual learners to create their own meaning (Jonassen et al. 1995). Traditional methods that evaluate learning within the four walls of a classroom do not transfer well to the online context. Evaluators need to understand the features of online activity and its social and ecological structure so that they may adopt new principles appropriate for evaluating learning. Typical tools for assessing beliefs and attitudes use Likert scale instruments: in the statistics education literature, there are only few such instruments. Gal and Ginsburg (1994) provide a survey, and point out some of their drawbacks. These include (a) inability to distinguish if the observed attitudes are actually towards mathematics or statistics, (b) inability to inform about the motivation behind attitudes unless student comments are included, and (c) difficulty identifying diverse attitudes using an instrument that measures only one or two dimensions of attitudes. In the light of these problems, some authors recommend the incorporation of qualitative methodologies (Lee 1999).

In a review of recent literature on online learning, we found that the majority of studies focused on the outcomes of learning rather than the process of learning (Cumming *et al.* 2002; Gudzial and Turn 2000; Hara and Kling 1999; Hendrinks and Maor 2003). Consequently, quantitative analysis predominates over interpretive inquires. While quantitative methods are often appropriate for investigating the medium of learning (Hara et al. 2000) or student characteristics, qualitative methods can provide more insight into the process of learning, including the quality of interaction and the relationship between interaction and knowledge construction (Gunawardena et al. 1997). The present study takes this approach, and seeks to identify important themes and provide in-depth discussion of students' views of online learning.

Methods

The theoretical framework for this study was the notion of *Community of Inquiry* established by Garrison and Anderson (2003). In this model, deep and meaningful learning, the central goal of higher education, takes place within a *Community of Inquiry* composed of teachers and learners as the key participants in the educational process. The model assumes that within this community, learning occurs through the interaction of three components: cognitive presence, social presence and teaching presence. The first element in the model is the development of *cognitive presence*, which Garrison *et al.* (2000) define as "the extent to which the participants in any particular configuration of community of inquiry are able to construct meaning through sustained communication." The second element is *teaching presence*, which includes designing and managing learning sequences, providing subject matter expertise and facilitating active learning. The third element is *social presence*, defined as the ability of learners to project themselves socially and emotionally in a community of inquiry. The function of this element is to support the cognitive domain through its ability to instigate, sustain, and support critical thinking within a community of learners. It supports affective objectives by making group interactions appealing and rewarding, leading to an increase in academic, social and institutional integration (and increased course completions).

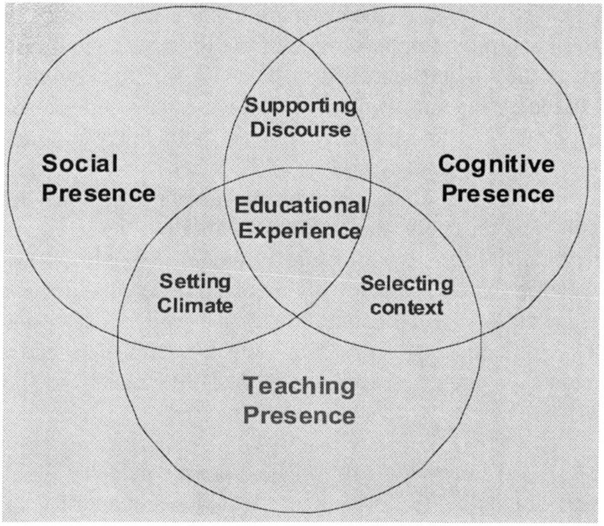

Figure 1: Elements of an educational experience (Garrison and Anderson 2003)

There were approximately 1,000 students enrolled in this subject and of these 269 volunteered to participate in the research. The study was approved by the appropriate ethics committees and participating students gave informed consent. The research group included students studying in two modes, online and traditional, and in two locations, campus-based and distance. More details about the participants are given in Suanpang et al. (2003).

The online course was developed into the *Electronic Classroom* type of online learning model, which engages the students in a planned series of activities via the Internet. In terms of Weller's (2002) *Learning Pedagogies*, this study combined constructivism, resource-based and collabourative learning. Due to limited time being available for constructing contents for the online course, only lectures 1-3 (Descriptive Statistics) and 6-8 (Inferential Statistics) were included, the remainder of the course being taught as normal. The *Learning Activities* planned included weekly study, weekly activity (virtual classroom discussion) and assessments. Online students had to participate weekly to complete their tasks with the minimum requirement of at least one hour per week. Communication between students and teachers was via telephone, e-mail, discussion board, and face-to-face by appointment. The traditional course was run using standard lectures, tutorials and homework assignments: the students in this group were also given a list of useful library resources to help them in their study.

The research was conducted in two phases, an orientation period followed by the main study. The purpose of the orientation phase was to introduce the online learning system and provide an example of the online course to the students. It was given to *all* the students enrolled in *Business Statistics* by electronic-based delivery (E-based). Students were notified of the web address for the course (http://e-learning.dusit.ac.th) and shown how to logon to the system.

Data were collected over a period of 16 weeks and the results from the online learning were then compared with those from traditional teaching. It is widely agreed that multiple methods in qualitative research are useful in achieving great understanding (Denzin and Lincoln 1994). It was with this mind that data collection for the present study involved different sources and kinds of information including conferencing transcripts, end of semester evaluation questionnaires, reflective stories from selected students, and open-ended questions. However, this research study had four main methods of data collection:

Interviews: Short interviews (10 minutes) were conducted with 20 campus-based students (the top ten and the lowest ten students in both the traditional and online groups) and 2 distance learning students (one traditional and one online). Group interviews were unstructured and took 30 minutes which not including students who had individual interview. The interview themes focused on experience of learning online course, likes and dislikes, problems and difficulties, use of online to support learning, and suggestions for further improvements.

Student diaries: Four students in the online group volunteered to keep weekly diaries of their learning and other study experiences. Students wrote about what happened in their learning, how they felt, what they liked and did not like, and other reflections and opinions.

Messages from discussion board: The discussion board was the "online community" where students posted their ideas, opinions, discussion, and comments reflecting on their learning. Several discussion topics were included in the online community ("Icebreaker", "What do you think of individual versus group projects?", "Experience of learning in an online course").

End-of-semester evaluation: There were three open-ended questions in the end of semester evaluation questionnaire. The purpose of evaluation was to give students the opportunity to express their feelings about their experience as well as give critical feedback of the strengths and weakness of the course. These questions were included "what students like and did not like about the course, experience of learning and comments for further improving.

To begin data analysis procedures, the interview recordings were transcribed and messages from the discussion board were downloaded: all the data were in Thai and required translation. The transcripts were read in their entirety, in order to obtain on overall sense of the interaction amongst students. This was followed this by multiple readings with a search for regular episodes of events, situational factors, circumstances, strategies, interactions and phases relating to the social construction of knowledge. The main focus of the analysis was on the "experience of learning", and the recurring regularities became the themes into which subsequent items were sorted (Hendriks 2003).

Results

Cognitive Presence

The first element in the model that is basic to success in higher education is "cognitive presence". Garrison (1997) defined this to mean the extent to which the participants in any particular community of inquiry are able to construct meaning through sustained communication. This is particularly important when the medium of communication changes, as in adoption of online learning for education purposes. Cognitive presence is a vital element in critical thinking, frequently presented as the goal of all higher education (Garrison 2003).

Theme 1: Students' background and prior knowledge

These are significant to construction of knowledge during interactions in the learning community. The result is a diverse pool of knowledge (Hendriks and Maor, 2003). Students who had good background in mathematics found that this subject was not difficult to learn. In contrast, students who did not have good background in mathematics found the subject difficult, and had negative attitude before entering the course. One recommendation from students was that the university provide an intensive course in fundamental mathematics and computing.

TradLow2: "The weakness is that it is difficult subject. Some students, who had no background in this subject or math before, would be afraid of it. It was like a wall so teacher had to overcome their feeling. We had bad impressed on our past experience because teacher did not care our feeling."

Themes 2: Students' perceptions of statistical learning context

This theme focused on how students understood Business Statistics, their ability to apply the most important statistical ideas and their attitudes to the subject. In general, students in the traditional classes seemed to understand and learning statistics by practicing and re-doing problems. Some lacked an appreciation of statistics, were not familiar with aspects of the subject, and held a view of statistics as a collection of isolated and meaningless techniques. For example, one student found that statistics was boring and blamed the way the course was delivered:

TradLow10: "I hate to study mathematics because there were a lot of numbers and formulas and I might get a headache trying to memorize everything. The more I try to understand, the more dizzy I get. The university should introduce new ways of learning maths to make students learning fun and enjoyable. Every time I go to class, I feel like I'm going to jail."

In contrast, online students had more experience with searching external sources, collecting, manipulating and representing data. Instead of spending most of their time learning and memorizing formulae and techniques, students were given the chance to conduct real investigations and gain experience of the wide applicability of statistical methods. A typical comment is:

OnlineTop1: "There were two types of experience, one was direct, other was indirect experience. Direct experience was knowledge of statistics, such as how to calculate statistics values from the real world case study. It was useful and could be used in my daily life. Statistics was not difficult and could be learned. Indirect experience was consciousness and awareness all the time when doing calculation. These two things were good. The last thing, I was proud of doing a good project. It made me eager to find out and continue to do."

Engagement with interesting activities through the use online course seems to have led to an increased enjoyment and appreciation of the subject. Students reported that the course covered much more than what they had thought. Before taking the subject, some of them had negative attitudes and they ended up finding the course quite interesting and useful.

Theme 3: Reflective and higher-order thinking

It is important to develop students' ability to evaluate their own work and challenge themselves. Students should develop their level of self-guidance and independent learning. It is import for students to learn to identify the components of effective learning and to be able to apply statistical concepts in their daily life. These abilities were particularly important for mature students in distance learning who are seeking to apply this knowledge in their careers. For example, one distance online student made this comment:

DistOnlineTop1: "I think the knowledge and experience that was gained from learning this subject could be used in nursing patients or at least it should be used in our daily life. I am in charge of nurses, so I expect that I should transfer this knowledge to younger nurses."

Social Presence

This is defined as the ability of participants to project their personal characteristics into the group, thereby presenting themselves to other participants as part of a learning community. This element is primarily a support for cognitive presence, facilitating the critical thinking carried on by the community of learners. When there are affective educational goals, as well as purely cognitive ones, then social presence contributes directly to the success of the educational experience (Garrison and Anderson 2003).

Theme 4: Communication in the learning environment

The most important factor in social presence is the communication and interaction in the learning environment. Communication among students and with teachers should effectively assist students to develop critical and high-order thinking. In the traditional model, Thai students are used to authoritarian practice, and are willing to accept what their teachers say without question, continuing the passive approach of their traditional schooling. Communication in this traditional mode was face-to-face, and few students were willing to ask questions of teachers. Online students gained advantages from using synchronous (chat) and asynchronous (discussion board and e-mail) communication methods. In particular, the discussion board opened up opportunities for shy students to reflect on their opinions and use it for communicating in learning and collabourative group working on their project, and students were happy to receive interactive feedback.

OnlineTop7: "It is good experience especially discussion board, it is better for communication than post message. We can exchange our ideas both learning and other topics. It makes someone who dears not to ask question in class, can show his idea or feeling such as the first experience in learning statistics forum."

Students also had comments about their experience related to technical problems in posting and editing messages in the discussion board. Some students remarked that shared reflection did not result in the level of interaction they would have liked, because some of their friends were slow in replying to messages.

Theme 5: Learning environment

Students commented that one important factor that influenced their learning was the class atmosphere. The formal nature of classroom environments can hinder learning, while the expansive nature of online interaction enhanced the learning environment and enabled students to construct personal knowledge. The following comment is typical:

GroupOnline1:"We always talk to friends in the chat room about content, homework, assignments and projects because sometimes the questions are very tough. We would rather discuss than copy like other subjects when we do not understand and just copy from friends. But in this subject it is different. We consult in our free time about the work, which we never did before, it's better than talking about F4 (the most favourite singer brand for Thai teenage)."

Theme 6: Collabourative learning

Collabourative group work was implement in the online mode only, as this environment offered opportunity for students to interact with others in a way that was not usual in the traditional mode. Students differed in their approaches to this new learning environment where there are fewer social cues to shape participation, and so the extent of interdependence and cooperation of each group varied. Cooperation was evident in some groups, while in others, each student took responsibility for the questions assigned to them and did not collabourate. There are interesting comments from students reflecting on their experience of collabourative group work:

OnlineTop2: "I got a lot of experience, some was good and some was bad. I had to adapt myself with members in the group and accept their opinions. I learned to divide work and include different opinions even through I did not agree with them. This made me know the difference between others and my thinking. It also gave me experience in human relationships that could be used in the future."

Teaching Presence

The third element of the model is teaching presence, an obvious function in higher education, whether traditional or online. In either case, it represents a means to support and enhance social and cognitive presence for the purpose of realizing educational outcomes (Garrison and Anderson 2003).

Theme 7: Course management and teaching technique

Students made many comments about the teaching techniques. Many in the traditional group had negative attitudes about learning by using the video-conference system, and complained about the high workload and the boredom of reading from the textbook. They suggested a reduction in the number of chapters studied, focusing only on content that was applicable to real life. Online students were satisfied with the course management and said that the teaching made statistics easier to understand:

OnlineTop3:"It depends on the method of teaching. The traditional teacher showed only slides so we had to study by ourselves. But our method of learning is new, such as studying by ourselves, doing individual and group projects and of course online learning. These made us have fun and become interested in learning, not bored because only theory could not be useful, like learning maths. I"ve got good experiences from this subject and I was impressed with the amazing teaching methods."

Theme 8: Technology

Students in the traditional groups suggested that they could use technology to assist their learning and reduce problems with using the textbook only. Online students benefited from the flexibility of being able to study "anywhere and any time," and discovered that the Internet could be used for educational as well as personal purposes:

OnlineTop6: "I used the online system to search for data in doing the individual and group projects. Before this, I used Internet for my pleasure such as chat, play online games and listen to music, rather than for learning. This online learning course made me know the benefits of using Internet for education. It also made me responsible to finish work on time and manage the extra time from the normal study access into online course without being forced."

However, some students also made negative comment about their experience related to technical problems with speed and reliability of the Internet.

Theme 9: Role of teachers

Students in the traditional groups claimed that the teacher was the most important factor influencing their learning. Online students also claimed that the teacher was an important factor in their learning, particularly in providing prompt feedback. Both groups of students were satisfied with the role of their teachers.

OnlineTop2: "Teacher should be nice, relaxed not too serious. She should not make us pressure like other subject that we had face such as Accounting, Economic and other subjects. These subject were difficult so teacher should be fund and attractive us better than teaching seriously. That would make us not understand and lack of motivation to learn."

Discussion

This study has explored and discussed students' learning experiences in the different modes of traditional and online learning in a Business Statistics course in the Thai context. In the traditional groups, it was found that students do most of their study during the lecture times, and carry out the exercises by themselves: however, they would benefit from increased access to technology in order to assist their learning. In the online groups, students used the online mode for reading course materials, doing exercises, searching for information, taking quizzes, and communicating with other students and online instructors. The way technology is currently being used in this Business Statistics course is to enhance the teaching of traditional material by presenting it in more exciting and more insightful ways.

Garrison and Anderson's (2003) model of an educational community of enquiry was used to organise the themes in students' comments on the course. The "cognitive presence" element found that students' background and prior knowledge are significant aspects for constructing knowledge during their interaction in the learning environment. Online students seemed to be more able to comprehend the statistics context, to apply statistical ideas to their daily life, and to evaluate and guide their own work. The "social presence" element found that online students had more effective communication (both asynchronous and synchronous) and higher level interactions than students in the traditional groups. Moreover, online collabourative learning offered the opportunities for students to share with others their thinking, interpretation, and reaction to the ideas that are being presented. Finally, the "teaching presence" element found that students in both the online and traditional groups were satisfied with the course management strategies and the role of their teachers. The teacher in the online course was more of a facilitator providing feedback, guiding learning and providing encouragement: the teacher in the traditional course had a more traditional and authoritarian role.

Despite its apparent advantages, the online course was not without problems. Without appropriate structures it could result in a sense of isolation and a lack of social interaction, and so it should include occasional face-to-face meetings for students to develop a learning community and social contact. Technical problems of access and speed of Internet need to be improved. In addition, the English language skills of Rajabhat students need to be improved, since English is required for accessing global information. Nevertheless, it seems that the use of such technology had a positive impact on the students' motivation and helped them gain better understanding and practice in their learning.

This study is one of the first research projects to investigate the use of technology such as e-learning in Thailand. Other aspects of the study are reported elsewhere: Suanpang, Petocz and Kalceff (2003) compare attitudes towards statistics between students in the traditional course and students studying using e-learning, Suanpang and Kalceff (2003) discuss features of the instructional design of such an e-learning system appropriate to the Thai context; Suanpang and Petocz (2003) discuss the improvements in scores of the online students as opposed to the students in the traditional groups. Overall, the results from this project are very positive and demonstrate the significant benefits of using e-learning to assist and support Thai students in their study of statistics. It seems that e-learning can become a powerful method for extending education in developing countries such as Thailand and other countries in South East Asia.

References

American Psychological Association. 1997. Learner-centered psychological principles: A framework for school redesign and reform, http://www.apa.org/ed/lcp.html (accessed February 9, 2003).

Berge, Z. L. 1997. Computer conferencing and the online classroom. *International Journal of Educational Technology* XXVII (11): 20-5.

Bonk, C. J., and Kim, K.A., eds. 1998. *Extending Sociocultural Theory to Adult Learning*. New Jersey: Mahway.

Cummings, J. A., Bonk, C. J. and Jacobs, F. R. 2002. Twenty-first century college syllabi: Options for online communication and interactivity. *The Internet and Higher Education* 5(1): 1-19.

Denzin N.K. and Lincoln, K. 1994. Introduction: Entering the field of qualitative research. In *Handbook of qualitative research,* eds. N. K. Denzin and K. Lincoln. Thousand Oaks, CA: Sage.

Gal, I., and L. Ginsburg. 1994. The role of beliefs and attitude toward science in school assessment and its use to investigate the relationship between science achievement and attitude toward science in school. *Journal of Research in Science Teaching* 25: 689-703.

Gardiner, L. F. 1997. Producing dramatic increase in student learning: Can we do it? *National Teaching and Learning Forum* 6(2): 8-10.

Garrison, D., and T. Anderson. 2003. *E-learning in the 21st Century: A Framework for Research and Practice*. London: RoutledgeFalmer.

Garrison, R., T. Anderson, and W. Archer. 2000. Critical inquiry in a text-based environment: computer conferencing in higher education. http://www.alt.ualberta.ca/cmc/publication.html (accessed October 15, 2003).

Gudzial, M., and J. Turns. 2000. Effective discussion through a computer-mediated anchored forum. *The Journal of Learning Science* 9(4): 437-69.

Gunawardena, C. N., C. A. Lowe, and T. Anderson. 1997. Analysis of an online global debate and the development of an interaction analysis model for examining social construction of knowledge in computer conferencing *Journal of Educational Computing Research* 17(4): 397-431.

Hara, N., and R. Kling. 1999. *Students' frustrations with web-based distance education courses*. http://firstmonday.org/issues/issue4_12/hara/index.htnml (accessed October 1, 2003).

Hara, N., C. J. Bonk, and C. Angeli. 2000. Content analysis of online discussion in an applied educational psychology course *Instructional Science* 28(2):

115-52.
Hendriks, V., and D. Maor. 2003. Qualitative methods in evaluating the quality of online learning. *World Conference on Educational Multimedia, Hypermedia and Telecommunication.* http://dll.aace.org/13307 (accessed July 9, 2003).
Jonassen, D., M. Davidson, M. Collins, J. Campbell, and B. Haag. 1995. Constructivism and computer-mediated communication in distance education. *American Journal of Distance Education* 9(2): 7-26.
Lee, C. 1999. A comparison of students' beliefs and attitude towards statistics between technology-rich environment and traditional lecture. *International Conference on Mathematics/Science and Technology* MSET 1(1), http://www.dl.aace.org/4797 (accessed June 10, 2003).
Matthew, D. 1999. The origins of distance educations and its use in the United States. *T.H.E. Journal* 27(2): 54-66.
Olson, M. T., and A. R. Wisher. 2002. The effectiveness of web-based instruction: an inquiry. *The International Review of Research in Open and Distance Learning* 3(2).
Richardson, C. J., and K. Swan. 2003. Examining social presence in an online course in relation to students' perceived learning and satisfaction. *Journal of Asynchronous Learning Networks* 7(1): 68-84.
Suanpang, P. and P. Petocz. (2003). Learning outcomes from a Business Statistics course in Thailand. *Fifth Annual Postgraduate Research Students' Conference,* University of Technology, Sydney, Australia. August 22, 2003.
Suanpang, P. and W. Kalceff. 2003. *Instructional design and features of an online Business Statistics course, Thailand.* The 29th Congress on Science and Technology of Thailand (STT). Khon Kean Thailand. October 20-22, 2003.
Suanpang, P., P. Petocz, and W. Kalceff. 2003. *Student attitude to Learning Business Statistics online Vs Traditional Method.* HERDSA 2003 Learning for an Unknown Future, Christchurch, New Zealand. July 3-6, 2003.
Weller, M. 2002. *Delivering Learning On The Net: The Why, What and How of Online Education.* British Library Cataloguing in Publication Data.

CHAPTER EIGHTEEN

COMPUTERS TEACHING ETHICS: KILLING THREE BIRDS WITH ONE STONE?

JOHN F HULPKE, AIDAN KELLY, AND MICHELLE TO
HONG KONG UNIVERSITY OF SCIENCE AND TECHNOLOGY

Computers Teaching Ethics: Killing Three Birds with One Stone?

Can ethics be taught? Put another way, how can students be led to think about ethics, especially in Asia where memorization is favored over thought? Secondly, how can sense of community, team learning, be fostered in big classes? Finally, how can ideas of individual participation be encouraged in Asia, especially in big classes? As in many places around Asia our university faced exactly these challenges and turned to computer technology for answers. To assess results it may be helpful to first review some frameworks to assess teaching ethics. Then a look at our classroom experiences may give insights into the three questions above.

Teaching Ethics: Frameworks

Measuring achievement in education is still an art more than a science. But Benjamin Bloom and colleagues have outlined a taxonomy of global educational goals (Bloom 1956) and Francis Reeves has looked at this work as a tool to help in teaching ethics. Bloom suggested grouping goals into three overarching categories, Cognitive, Affective, and Motor Skills. While teaching and learning Motor Skills has obvious relevance in brain surgery and piano playing it is not immediately apparent how Motor Skills relates to teaching ethics. But it may be useful to consider both cognitive and affective domains when looking at our issue, teaching ethics in universities, especially in Asia. Reeves explains how.

In education in general, and perhaps especially in Asia, traditionally COGNITIVE goals are those recognized as important. In teaching ethics affective factors may be even more important, but let us first consider the cognitive side. Bloom's first level in the cognitive area is *Knowledge.* For example, "name one firm studied in this class whose executives were sent to prison." This can be answered by students who memorize facts. Equity Funding is one possible answer, and a early classic in computer fraud. We also ask students to know seven ways to approach a question of ethics, and the seven happen to sat with the letters J, U, S, T, I, C, and E. Students can be tested to see whether they know, have the desired *knowledge,* relating to our JUSTICE model. The JUSTICE model is mentioned here in order to illustrate, so a lengthy discussion of the model is not needed here. But briefly, the model asks an individual to choose three of seven possible sets of ethical decision making criteria and apply the three to an issue to be decided, whether helping a neighbor cheat or reporting misconduct by your boss. In the JUSTICE model each letter stands for one approach:

JUSTICE, applying same rules to all fairly and evenly
UTILITARIAN thought, greatest good for greatest number, does good outweigh bad
SPIRITUAL VALUES, such as the Golden Rule, to do to others as you would want others to do to you
TV RULE, knowing you must explain your decision on TV with your family watching
INFLUENCE criterion, considering how big an influence (if any) your actions would have your
CORE values, the deepest human values, considering things REALLY important in life the urgency of the decision, whether this is a life or death issue, an
EMERGENCY requiring immediate action

Students in our classes seemed to have little trouble mastering the seven criteria, and having the knowledge to define each, even in a short one credit class.

The second cognitive level Bloom discusses, as interpreted by Reeves, has to do with *Comprehension.* Who were the stakeholders in the Enron case? What could a group of stakeholders or an individual stakeholder done to change the outcome of the case? When? How? So what? *Application* is a third cognitive level. The case method is often used in ethics courses. Students are typically given a case or mini-case and are expected to go though a series of steps to insure they can apply what is learned in class to the case in question. For example, one approach lists seven steps:

- Defining the issue
- Analyzing the case data with focus on causes and effects as well as constraints and opportunities
- Generating alternatives
- Selecting decision criteria
- Analyzing and evaluating alternatives
- Selecting the preferred alternative
- Developing an action and implementation plan

In our classes we showed a popular ethics video, The Parable of the Sadhu. In this video a Wall Street broker describes encountering a half dead monk or pilgrim, a holy man called a Sadhu, while in the middle of a long planned mountain climbing expedition in the Himalayas. Students were asked apply these or similar steps, showing they could apply what things they have learned (such as the JUSTICE model) to the situation in the video. After *Application* Bloom moves to a fourth cognitive level, *Analysis.* Reeves suggests that classroom discussions or written examinations can check skills of *Analysis* by asking about unstated assumptions, causality, or fallacies for example. A stage beyond *Analysis* is *Synthesis. Synthesis* may be asked for solutions to an ethics problem the class has identified as existing in the university (cheating in exams, plagiarism involving web, copying home work). New creative approaches to the issues identified suggest students are developing abilities to *Synthesize.* Bloom's sixth and final cognitive level is *Evaluation.* Reeves, citing Bloom, says *Evaluation* is "making a judgment about the value or worth of something. It presupposes the previous levels but is not necessarily the last step" (Reeves 1990. p613). In an ethics class students should be pushed to reflect, to evaluate, to carefully judge situations, look closely and critically at alternatives, and consider implications of outcomes. *Evaluation* is not easy but is important.

Although COGNITIVE factors are important in an ethics class, the AFFECTIVE domain looms even larger. Bloom talks about five affective levels. First he calls *Receiving.* Each student in the ethics class, provided he or she stays awake, will be exposed to ideas related to ethics. While there may be a more-or-less neutral reaction to the JUSTICE model, at least the student has *received* input, and should hopefully not feel negatively towards the idea of deciding ethically. A second level is called *responding.* Here Bloom is talking about doing assignments, either from fear of bad grades or out of interest, or perhaps engaging in class activities because they seem worthwhile, worth the time. Carrying our JUSTICE model example a bit further, students at the *responding* level may get some pleasure out of figuring how to apply the model to a case of student cheating for example. There is no burning desire to become an expert in ethics, but at this level there is a *responsive* feeling in the student. *Valuing* is the third level in Bloom's affective typology. Once a student feels that "ethics is an important topic" the student can be said to have reached the *Valuing* level. While a good classroom lecture might help a student reach the *received* level and also the *responding* level, Bloom suggests that the final three levels, including *valuing*, require participative learning. Lectures are not enough. Bloom also suggests *Organization* and *Characterization* can be seen as more affective than cognitive. Once a student can *organize* and internalize a value system, fore example think through and argue for or against a framework or model such as the JUSTICE model, that student can be said to have reached the *Organization* affective level. Bloom calls his fifth and final affective level *Characterization.* Although one might wish Bloom had chosen a more affective-sounding term, as explained *Characterization* can be used to describe an individual who as reached a high stage in the affective domain. If a person can be characterized as an exceptionally helpful person, or uncommonly honest, then enough values must have been internalized so that the characterization fits the individual. That IS the person, perhaps the HONEST person Diogenes looked for.

Bloom's model, says Reeves, can be used in any discipline, to cover any type of learning. Reeves contribution is to show that indeed Bloom's taxonomy can be applied to teaching or learning ethics. Later we can look again at Bloom's six cognitive levels and his five affective levels, and check how computers may or may not be useful in helping students attain these levels.

While Bloom's discussion focuses attention on the learner, it may be useful to also look at a framework suggested for looking at CONTENT of an ethics course. Dianne Martin, Chuck Huff, and several colleagues have suggested such a framework in a series of papers. One graphic used in these papers is shown here (Huff and Martin 1995, 76):

Technologies

		Topics of Ethical Analysis								
		Responsibility		Ethical Issues						
		Individual	Professional	Quality Of Life	Use of Power	Risks and Reliability	Property Rights	Privacy	Equity and Access	Honesty and Deception
Levels of Social Analysis	Individuals									
	Communities and Groups									
	Organizations									
	Cultures									
	Institutional Sectors									
	Nations									
	Global									

Figure 1. Framework for the content of an ethics course (Huff and Martin 1995, 76)

Bloom's taxonomy is intended to be applicable to any discipline. The Martin and Huff framework is aimed squarely at teaching ethics. While any number of topics might be covered in an ethics class, the value of the Huff-Martin framework is that it lists some commonly covered topical areas (privacy for example) and can thus be used to check one's own course. In addition, the Huff-Martin framework shows seven levels of analysis ranging from individual to global. Again, such a listing of levels of analysis can help one spot ways to possibly improve a course.

It may be valuable to look at one specific ethics course at our university using both Bloom's taxonomy and Martin and Huff's framework. Then we can look at ways we tried or might have tried to use computers in the course. But first consider the background and circumstances of our specific course.

The Experimental Ethics Class

With this background in mind, a look at one case may suggest how theories may work or not work in practice. How did our efforts measure up against Bloom's taxonomy? Bloom's first level in the cognitive area is *Knowledge.* Certainly our course met objectives here. Two quizzes helped assess acquisition of knowledge. The quizzes may have also served as a motivating force to remind students to study in order to obtain important facts (and to retain these facts at least until exam time). Bloom's second cognitive level is *Comprehension.* Quizzes were essay not multiple choice, so *Comprehension* could also be encouraged and assessed. *Application,* Bloom's third cognitive level, can also be evaluated by essay questions. But here other elements in the class were of more direct value. In a course with limited classroom time it would be impossible to involve all in arguments or debates. But debates are an excellent way to have students apply concepts covered in lectures. The solution here was to have teams engage in on-line debates, with each side required to post at least three entries. First the proponents had to make a statement, then the opposing team had to reply, then at least two more iterations took place. After *Application* Bloom suggested a stage he called *Analysis.* Students can be asked about unstated assumptions, causality, or fallacies for example. But how can this be done with 300 or more students in the room? This is where our university's Personal Response System (PRS) came into play. Our university wired all classrooms computer based "PRS" technology, a Personal Response System where all students can "vote" real time in the classroom. We posed ethical dilemmas by written caselets, dilemmas described in class, and/or short videos. Then students were shown 4 or 5 alternative choices and the 30 seconds to vote. All students voted, results were displayed, followed by analysis and discussion. For example, in the video The Parable of the Sadhu, we pause in the middle of the video and ask "what steps do you think the Wall Street broker took in this situation?" We give choices, ranging from "give the stranded half dead Sadhu coffee and say good bye" to "take the Sadhu to safety even though it may mean abandoning the long planned mountain climbing expedition." Students almost unanimously predict that the broker would take the Sadhu to safety, but in real life this did not happen. The Sadhu was passed off to another group of mountain climbers with no in likelihood of anyone ever saving the Sadhu. This leads to a great awakening on the part of students almost all of whom had pressed the wrong response on the PRS handset, and each now individually knew that his or her unstated assumptions were incorrect.

A stage beyond *Analysis* is *Synthesis. Synthesis* may be asked for solutions to propose a plan to solve a real ethical issue. But how can more than 300 students do this? It cannot be done in the limited class time available. But by requiring students to respond individually to a dilemma posted on the web, each moves to the level of *Synthesis.* One such example involved what a patrol team should do when a boatload of people try to illegally enter the area. An element of uncertainty and possible danger made this issue very messy, requiring synthesis to handle. Other similar mini-cases were used also. The *evaluation* level is the highest level in Bloom's taxonomy of cognitive skills. The on-line debates required each team to evaluate the assumptions and arguments of the opposing team. Again, without the use of computers and the web in particular, it is hard to imagine pushing students to this level in a short 8 hour course typically with 300 students in the classroom.

But debates did more than this: In almost every case, students responded to their "opposing team" emotionally. Clearly affective dimensions came into play here. The levels of *receiving* and *responding* can be tested in a big class using the PRS system. But the debates clearly moved past those levels, as values were stated and defended time after time in debates. Bloom characterizes *Organization* as "organization of a value system" and in short essays each student posted to the web, *organization* became apparent. For example, in one class, each student had to describe his or her own view about Corporate Social Responsibility. They did the assignment and were often quite emotional in supporting their positions. As such, most probably reached what Bloom called *characterization,* as responses showed what kind of person each student was deep inside. Could this have been done without the computer and the web? Yes, in a smaller class, but probably not in a class the size we faced.

Knowledge	learn facts: memorize a decision making model
Comprehension	news item: who are stakeholders?
Application	Sadhu: go through case steps, cause effect etc
Analysis	Sadhu: causality (bad examples, etc), fallacies
Synthesis	exam cheating: issue raised by class, then solutions proposed
Evaluation	plagiarism using WWW: problem or not? Culturally defined?

Figure 1. Bloom's learning model applied to ethics class: Cognitive domain

Receiving	Hear lecture about decision making model, watch corruption video
Responding	Apply decision making model to genetically modified food
Valuing	Students decide course topic important to life success, national success
Organization	Students think through and modify lecture model, make own model
Characterization	Students decide "this is me, important to me, I will do the right thing"

Figure 2. Bloom's learning model applied to ethics class: Affective domain

Another way to assess results of our course might be to look at the Martin-Huff framework. Did the course address issues ranging from the individual to the global level of analysis? It did, from the JUSTICE framework for individual decision making to web discussions of global warming. Did our course cover topical issues shown on the Martin-Huff framework, ranging from quality of life to privacy? Not really, or at least not in an organized way. Quality of life entered into the course when teams of students were required to prepare and post web pages on "good news about the environment" or "bad news about the environment." Web discussions and quiz questions about issues such as child labour and sexual harassment. But in a one credit course, it seemed better to go for at least a minimum amount of depth rather than insure coverage of all relevant possible topics.

Finally, let us return to the original three questions posed at the beginning of this essay. First, can ethics be taught? Or phrased another way, can Asian students be led to think about ethics? Second, can a sense of community be built in classes of 200 and more students? Finally, can individuals participate in such big classes? And for all three questions, can computers help? A look at our classroom experiences may give insights into these three questions.

First, can ethics be taught? Can students be led to think about ethics? While there is considerable disagreement as to the answer to these questions, there is less disagreement on a more important question: can ethics be LEARNED? Most would say yes. But this still leaves academics with the job of facilitating that learning. And many have noted that it is not easy to set the stage so students learn ethics. In fact, some find it easier to bypass the question, and simply TEACH ethics, hoping some learn. While this issue troubles teachers of ethics world-wide, there are special cultural features which suggest that feasible solutions may be a bit different in the Asia Pacific region. Many would agree that the very essence of any study of ethics involves thought. Thoughtful reflection is part of what ethics professors hope students learn. Yet, education in China, and in many parts of Asia, stresses individual rote learning. Listen, memorize, recite. This emphasis has caused by geographic region to be labeled as having a "test culture." Indeed where we teach many students cannot get past the idea they have held for years. "Is this going to be on the exam?" If not, expect most students to tune out. While this is more prevalent in heavier subjects such as ethics compared to mechanical engineering, one still hears the question in technical fields too. Thinking is not a goal in many student minds. Students all too often are stuck at Bloom's first level in the cognitive area, *Knowledge.* They may master facts, obtain knowledge, but how can we get our Asian university students to think? And as others have often pointed out, university students should not be expected to immediately embrace the study of ethics outside of Asia either. Issues raised in ethics courses are complex and not easy to handle (Cooper et al, 2001).

So, we are faced with the first of our unanswered questions, how to best promote thinking and to best facilitate learning of ethics. And, especially in the Asia Pacific region, how best to "teach" ethics, given traditional emphasis on memorization?

With this first question still on our minds, let us look at our second and third issues. Can a sense of community, team learning, be fostered in big classes? And can individual participation be promoted, something many say Asian students sorely need? One might try to avoid or evade these questions by decreeing that ethics shall be taught in small classes. This sounds good, but a number of disparate sources put pressure which resulted in our facing 175 to 300 or even more faces each time we tried to "teach" ethics. First, when ethics is an elective course, there is a potential for those who most "need" the course to opt to not take the course. This is an especially troubling thought in the Business School, where arguably there is a crying need for some sort of ethics exposure. So, we argued that ethics should be required. But business school faculty are not especially fond of adding new required courses, and are surely not inclined to add a required course in ETHICS. After much fighting and compromising, our university was convinced to add a required ethics course for ALL business undergraduates… but on a couple of bothersome conditions. First, it could only be a one credit course. Second, one of the proposers of the course would have to agree to teach ALL the students, some 750 students each year. (In addition, it was assumed that we could continue to teach other courses, as this was only a one credit course!)

Let's take the "team" and "participation" issues one at a time. With 300 students in our lecture hall, can we foster a sense of community, encourage team learning? Teams are possible even with a large number of students, but the "one credit" provision meant we only had students in our lecture hall about 8 hours total. Form and build teams in eight hours?

And individual participation did not seem practical given the constraints faced. If each of our students asked a 30 second question, a worthy goal, and we gave a 60 second answer to each, we would have used up most if not all of the time we enjoyed in the presence of our students.

So, three issues: First, how can ethics best be learned, especially where memorization is favored over thought? Second, how can team learning be fostered in a short period of time? Finally, how can individual participation be encouraged in Asia in big classes?

In each case, technology provided possible help.

First, thought instead of memorization of facts? Two things were tried. First, as mentioned above our university had equipped our classroom with "PRS" technology, a Personal Response System where students "vote" real time on issues or simply to chose a statement closets to his or her personal opinion. We posed ethical dilemmas by written caselets, dilemmas described in class, and/or short videos. Then students were shown 4 or 5 alternative choices and the 30 seconds to vote. All students voted, results were displayed, followed by analysis and discussion. In addition to using PRS to encourage thought, each student individually had to respond on a class discussion board to 5 ethical dilemmas posted on the web. Finally, teams (formed and solidified by technology as will be shown later) had to engage in week-long debates on the web. Each team of four or five students participated on one debate out of about twenty that went on simultaneously. Thinking, even critical thinking, was apparent. Others have described many ways technology can enhance learning of ethics (Tavani 1998; Golden, et al. 2001) but we found technological tools especially useful in a big class. In an Asian context where participation is not the cultural norm Hongladarom suggests "Active participation by students should be encouraged. The internet greatly facilitates this" (Hongladarom 2001). So, both culture and large class size suggest using technology in teaching ethics.

Second, how to form and build teams given time constraints? Preplanning and lots of work was a key here. All students were assigned to teams from A to Z before class began, and assigned seats with their teams. All seats had big name tags also, facilitating name recognition. Within a few minutes of class beginning, each team was asked to decide on a team name, with the name beginning with the same letter of the alphabet assigned in advance. Thus team A may call themselves Acrobats and team P may become Pandas, etc. Further, in a very short time, each team spread around the floor and aisles and prepared a flip chart depicting a team "logo," a poster which included the team name, the name of each student on the team, plus artwork of some kind (a Panda for example). Each team taped their poster on the auditorium wall (carefully, with paint-safe masking tape). Photos were taken of each team next to its poster. The after a bit of lecture and case work, using the PRS, each team was assigned the task of going to a computer barn on campus and in 30 minutes finding and bringing back an article about corruption in our country. While teams were doing this computer lab work, the course teaching assistant feverishly loaded each team photo onto the class web site, without announcing this. An amazing thing happened each time. At least one student would look at the class web page while working on the team assignment on corruption, and would notice the team photos. Word would spread through the computer barn like wildfire. Soon all teams wanted to see THEIR team photo on the web! Within two hours after class began, groups had formed into TEAMS.

Finally, how can individual participation be encouraged in a large class? Again, the computer helped. Each student individually responded on a class discussion board to 5 ethical dilemmas. The dilemmas were not easy, and answers required thought. There is no way 750 students could express their opinion out loud in class on even ONE ethical issue, but now each of the 750 wrote short essays defending their position on five dilemmas.

Was this a success, using one stone, the computer, to kill three birds? Pretty well, but as one could imagine, it took time; in fact, we almost died of overwork. After the end of the course in a presentation at the university about new learning methods we summarized results as follows: "Great success. But never again." In fact, the temptation was to return to what has been the educational norm in too much of Asia: "Hello students: sit down, shut up and listen!" Although we have not gone quite so far, most of the innovations described above have been abandoned. Were our efforts a success, or perhaps a "qualified success?" We think it would be better to call it a qualified failure. Answers are not easy to see. All three problems still bother us. Hopefully others can take the ideas presented here and make them workable.

References

Agarwal, J. and D. C. Malloy. 2002. An integrated model of ethical decision-making: a proposed pedagogical framework. *Teaching Business Ethics* 6(2):245-268.

Bloom, Benjamin S. 1956. *Taxonomy of Educational Objectives*. New York: David Mackay Company, Inc.

Bruton, Samuel V. Teaching the golden rule. *Journal of Business Ethics* 49(2): 179-187.

Cooper, Gregory, David Elmes, and Jeanine Stewart. 2001. Issues in teaching ethics to undergraduates. *Council on Undergraduate Research Quarterly* Dec 2001: 55-56.

Giacalone, R. A. 2004. A transcendent business education for the 21st century. *Academy of Management* 3(4):415-420.

Golden, Ilya, Kevin Ashley, and Rosa Pincus. 2001. *Introducing PETE: Computer Support for Teaching Ethics.* http://pete.lrdc.pitt.edu.

Gotterbarn, Don, Chuck Huff, Dianne Martin, and Keith Miller. 1996. A framework for implementing and teaching the social and ethical impact of computing. *Education and Information Technologies 1(1).* Published by Chapman, http://onlineethics.org/topics/accountability.html

Hongladarom, Soraj. 2004. *Cultural issues pertaining to the use of internet in teaching and learning in Thailand.*: http://pioneer.netserv.chula.ac.th/~hsoraj/web/soraj.html.

Hooker, John. 2003. *Working Across Cultures*. Stanford, CA: Stanford Business Books.

Huff, Chuck and Dianne Martin. 1995. Computing consequences: a framework for teaching ethical computing. *Association for Computing Machinery: Communications of the ACM* 38(12): 75-85.

Morgan, Eileen, 1998. *Navigating Cross-Cultural Ethics: What Global Managers Do Right to Keep from Going Wrong*. Butterworth-Heinemann.

Piper, Thomas R, Mary C. Gentile, and Sharon Daloz Parks. 1993. *Can Ethics*

Be Taught? Boston: Harvard Business School.

Rawls, John. 1998. *Distributive justice.* In *Perspectives in Business Ethics,* ed. Laura Pincus Hartman, 44-53. Chicago: Irwin McGraw Hill.

Reeves, M. Francis. 1990. An application of Bloom's taxonomy to the teaching of business ethics. *Journal of Business Ethics* 9(7): 609-616.

Rossouw. G. J. 2002. Three approaches to teaching business ethics. *Teaching Business Ethics* 6(4): 411-433.

Tavani, Herman. 1998. Internet resources for teaching computer ethics. *APA Newsletter* 97(2), www.rivier.edu/staff/httavani/.

Terri, L. H., and D. L. Gilbertson. Ethical principles vs. ethical rules: the moderating effect of moral development on audit independence judgments. *Business Ethics Quarterly* 14(3): 499-523.

Trevino, L. K., and K. A. Nelson. 1999. *Managing Business Ethics: Straight Talk about How to Do It Right*. John Wiley & Sons.

Woolsey, G. 1974. Homage to W. A. C. Bennette and the Canadian Operations Research Society. *Interfaces* 4(4): 43-46.

CONTRIBUTORS

Darren Abramson received his PhD in Philosophy and Cognitive Science from Indiana University in 2006, with a Certificate in Logic. He also holds a Masters degree in Computer Science from Indiana University. Currently he is an assistant professor at Dalhousie University where he teaches Computer Ethics, Logic, Cognitive Science, and Philosophy of Psychology. His current research interests include computational simulations of the cortex, ethical implications of specification and verification, and computational creativity. *Contact information:* Darren Abramson; da@dal.ca; Department of Philosophy, Dalhousie University, Halifax N.S. Canada B3H 4P9.

Yeslam Al-Saggaf is a lecturer in the School of Information Studies and a Course Coordinator for the Bachelor of Information Technology Partnerships) at Charles Sturt University, Australia. He is also a Research Fellow in the Centre for Applied Philosophy and Public Ethics CAPPE. He holds a bachelor's degree in engineering (with honours) in computer and information engineering, from Malaysia, and a master of information technology and a PhD from Charles Sturt University, Australia. His research interests lie in the areas of online communities (both social and political). He has published in those areas in a number of international refereed journals.

Michael Byron is Associate Professor of Philosophy at Kent State University in Kent, Ohio, where he has held an appointment since 1997. He is the editor of *Satisficing and Maximizing: Moral Theorists on Practical Reason* (Cambridge University Press, 2004), the co- author (with Deborah Barnbaum) of *Research Ethics: Text and Readings* (Prentice-Hall, 2001), and the author of numerous journal articles in ethical theory and theory of rationality. During 2004–05, Dr. Byron was Visiting Fulbright Scholar in Philosophy at Lingnan University in Hong Kong.

Robert Cavalier received his BA from New York University and a Ph.D. in Philosophy from Duquesne University. In 1987 he joined the staff at Carnegie Mellon's Center for Design of Educational Computing (CDEC), where he became Executive Director in 1991. While at CDEC, he was also co-principal in the 1989 EDUCOM award winner for Best Humanities Software (published in 1996 by Routledge as *A Right to Die? The Dax Cowart Case*). He also co-authored the *CD-ROM The Issue of Abortion in America* (Rountledge, 1998) Dr. Cavalier is currently Director of CMU's Center for the Advancement of Applied Ethics and Political Philosophy. He is also Director of the Center's Digital Media Lab which houses Project PICOLA (Public Informed Citizen Online Assembly) and co-Director of Southwestern Pennsylvania Program for Deliberative Democracy. He is co-editor of *Ethics in the History of Western Philosophy* (St. Martin's/Macmillan, England, 1990), editor of *The Impact of the Internet on Our Moral Lives* (SUNY, 2003) and other works in ethics as well as articles in educational computing, Dr. Cavalier is internationally recognized for his work in education and interactive multimedia. He was President of the "International Association for Computing and Philosophy" (2001 - 2004) and Chair of the APA Committee on Philosophy and Computers (2000-2003).

Ludmil Duridanov is a lecturer at the Institute of Slavic Studies, Friedrich-Schiller University of Jena. He obtained is Dr. phil. degree on general linguistics, slavic studies and philosophy from Albert-Ludwigs-University Freiburg. Right now he is engaged in a research project, with Simeon Simoff, on virtual reality as the foundational concept of hearing since 2001. Among his publications are *Der Text als erzeugende Prinzip von byzantisch-slawischen Paradigmen*, *Ein Paradox des Ikonoklasmus: das ursprüngliche Bedeutungsfeld von GRAPHEIN*, and *Über die Kommunikation als philosophisches Problem*.

Charles Ess is Professor of Philosophy and Religion and Distinguished Professor of Interdisciplinary Studies, Drury University (Springfield, Missouri, USA); and Professor II, Programme for Applied Ethics, Norwegian University of Science and Technology (Trondheim). Dr. Ess has received awards for teaching excellence and scholarship, and published extensively in comparative (East-West) philosophy, applied ethics, discourse ethics, history of philosophy, feminist Biblical studies, and Computer-Mediated Communication (CMC). With Fay Sudweeks, Dr. Ess co-chairs the biennial conferences "Cultural Attitudes towards Technology and Communication" (CATaC), and has served as a Visiting Professor, IT-University, Copenhagen (2003), and a Fulbright Senior Scholar, University of Trier (2004). Contact information: Interdisciplinary Studies Center, Drury University, 900 N. Benton Ave., Springfield, MO 65802 USA. Homepage: http://www.drury.edu/ess/ess.html.

Alberto Gatti graduated in Philosophy in 1999 at the University of Pavia (Italy) and received the PhD in Cognitive Science in 2004 at the University of Siena (Italy). He has dealt with the issue of representations in cognition and, in particular, with the function of representations in the processes of information elaboration and with the role of human-environment interaction in generating the representations themselves. He has published papers about the specific epistemic role that certain types of actions performed by humans upon the environment reveals. Today he works at an Italian company that makes research and development in the field of eye-tracking technology.

Ken Herold is Director of Library Information Systems at Hamilton College in New York. Herold earned his Master's degree in Library and Information Studies from the University of California at Berkeley in 1984. In 2002, he was appointed as a Correspondent to Oxford University Computing Laboratory's Information Ethics Group. He participated as a senior practitioner in a recent international Delphi study of the foundations of Information Science. Herold also serves on the editorial board of *Library Philosophy and Practice* (ISSN 1522-0222).

Lawrence M. Hinman, Ph.D., is Professor of Philosophy and Director of the Values Institute at the University of San Diego. He is also Co-Director of the San Diego Center for Ethics in Science & Technology. He is the author of numerous articles in philosophy as well as *Ethics: A Pluralistic Approach* (Thomason/Wadsworth) and *Contemporary Moral Issues* (Prentice-Hall), both in their third editions. He is also the founder and editor of Ethics Updates (http://ethics.sandiego.edu). Most recently, he has co-edited an issue of *Stem Cell Reviews* devoted to the ethics of stem cell research.

Soraj Hongladarom is an associate professor of philosophy and Director of the Center for Ethics of Science and Technology (http://www.stc.arts.chula.ac.th/) at Chulalongkorn University in Bangkok, Thailand. He has published books and articles on such diverse issues as bioethics, computer ethics, and the roles that science and technology play in the culture of developing countries. His concern is mainly on how science and technology can be integrated into the life-world of the people in the so-called Third World countries, and what kind of ethical considerations can be obtained from such relation. He is co-editor, with Charles Ess, of *Information Technology Ethics: Cultural Perspectives,* forthcoming from Idea Group Publisher. *Contact information:* Soraj Hongladarom, Department of Philosophy, Faculty of Arts, Chulalongkorn University, Bangkok 10330, Thailand; hsoraj@chula.ac.th.

John Hulpke has held positions in government and the private sector, including the U S Post Office, truck driver, shipyard electrician, US Air Force Officer, Visa Officer for the US Department of State. After returning to school to earn a Ph D (University of California, Berkeley) Hulpke taught business in Ireland, Korea, Mexico, and the USA. He joined Hong Kong University of Science and Technology (HKUST) in 1997, where he teaches management. In 2001 and 2002 Hulpke was MBA Director at the China Europe International Business School (CEIBS) in Shanghai, but is now again full time at HKUST.

Seungkwon Jang received his Ph.D from Lancaster University Management School in U.K., and is Associate Professor at Sungkonghoe University in Seoul, Korea, and Honorary Research Fellow at the University of Aberdeen Business School in U.K. He worked previously at Samsung Economic Research Institute as Chief Research Fellow. He is the author of books and research papers on postmodern organization theory, complexity theory, and social aspects of information communication technologies. His main research interests focus on the conceptual approaches to organizational and technological innovation of information communication technologies.

Aidan Kelly is Professor of Management at the Smurfit Graduate School of Business, University College Dublin. He makes frequent research trips to Asia, and is interested in changing career development patterns. He continually supervises doctoral students, many of whom focus on areas related to this paper. He contributes frequently to academic journals and conferences. Professor Kelly has had a distinguished administrative career, holding such posts as department head, dean, Vice President (Development). He now teaches full time, focusing on Organizational Behaviour and Organization Structure and Design, primarily at the graduate level. He has served as visiting scholar in China Europe International Business School (CEIBS) in Shanghai and at Hong Kong University of Science and Technology (HKUST).

Gerald Khoury has nearly twenty years experience in IT planning, architecture, development, consulting and management. Gerald has performed the roles of Architecture Manager, Chief Architect and Architecture Consultant for a variety of large corporations. This has given him a deep understanding of the needs of business and the way in which IT services can be positioned to provide strategic competitive advantage and strengthen vendor-client relationships.

Lorenzo Magnani, philosopher and cognitive scientist, is a professor at the University of Pavia, Italy, and the director of its Computational Philosophy Laboratory. He has taught at the Georgia Institute of Technology and at The City University of New York and currently directs international research programs in the EU, USA, and China. His book *Abduction, Reason, and Science* (New York, 2001) has become a well-respected work in the field of human cognition. In 1998, he started the series of International Conferences on Model-Based Reasoning (MBR). The last book *Morality in a Technological World: Knowlledge as a Duty* is in press with Cambridge University Press.

Aaron Quinn is Assistant Professor of Journalism at California State University, Chico. Aaron teaches and researches in media ethics, visual journalism, computing and technology, and applied philosophy generally. Aaron recently submitted his PhD Thesis, The Rehabilitation of Professional Journalism, out of the Centre for Applied Philosophy and Public Ethics in Canberra, Australia. Prior to entering academia Aaron was a photojournalist, reporter and editor at several U.S. Newspapers.

Simeon J. Simoff is a senior lecturer in computing science and information technology and one of the founders of the e-Markets research group at the Faculty of Information Technology, University of Technology, Sydney. He is known for his unique blend of interdisciplinary research, which integrates the areas of data mining, design computing, computer-mediated collaborative work and digital media. He is one of the founders of the leading international research and educational programs on virtual design studios, established in Australia and reflected in his recent co-authored monograph "Understanding Virtual Design Studios" (Springer-Verlag, 2000). Presently, he is an active member of several professional societies and groups. As a member of ACM SIGMOD and SIGKDD, Simeon has initiated and co-chaired annual workshops on multimedia data mining and on visual data mining. He is the Australian regional editor of the ASCE International Journal of Computing in Civil Engineering, and active member of ASCE Intelligent Systems Committee, and Data and Information Management Committee. He holds a PhD in Computer Science.

Pannee Suanpang is currently Director of Academic Resources Center and Information Technology. She is the program director of Master in Science of Knowledge Management and program director of Bachelor of Science major Information Technology, Suan Dusit Rajabhat University, Bangkok, Thailand. Her research has involved developing a pioneer online course in Thailand and studying the effectiveness of online learning for Thai students. Finally, she is doing several research projects including, the investigation and analysis of information and communication manpower of Thailand; the ICT manpower development and the development of the universities in Thailand to be learning organization by using knowledge management.

Chulanee Thianthai received her Ph.D. in Cultural Anthropology from the University of Oregon. She joined the Faculty of Political Science, Chulalongkorn University, Bangkok, Thailand, as a full time faculty member of the Department of Sociology and Anthropology in 1996. Her expertise is in the areas of cultural anthropology, medical and nutritional anthropology, business anthropology, gender issues, urban population, and teenager lifestyles. As an anthropologist, her research brings out viewpoints and in-depth understandings we often neglect to see. Her works will continue to fascinate us with aspects found in modern city and contemporary issues. Contact e-mail: chulane.t@chula.ac.th

Michelle To has wide ranging experience in local and international organizations including airlines, conference organizers, transportation, travel agency, FMCG, investment bank and information technology companies. She is a member of professional bodies including the Institute of Chartered Secretaries and Administrators in the United Kingdom, the Hong Kong Institute of Company Secretaries and the Hong Kong Association of Business and Professional Women. She is a Certified Hospitality Educator awarded by the American Hotel & Lodging Association. She currently works at the Department of Management of Organizations of the Hong Kong University of Science and Technology.

John Weckert is Professorial Fellow at the Centre for Applied Philosophy and Public Ethics, an Australian Research Council funded centre based at Charles Sturt University, The Australian National University and the University of Melbourne, and Professor of Information Technology in the School of Information Studies, Charles Sturt University. He has published widely in the ethics of information and communication technology and is the founding editor-in-chief of the Springer journal Nanoethics: Ethics for Technologies.